Quantum Measurements and Decoherence

Fundamental Theories of Physics

An International Book Series on The Fundamental Theories of Physics: Their Clarification, Development and Application

Volume 110

Quantum Measurements and Decoherence

Models and Phenomenology

by

Michael B. Mensky

P.N. Lebedev Physical Institute,
Russian Academy of Sciences,
Moscow, Russia

Springer-Science+Business Media, B.V.

A C.I.P. Catalogue record for this book is available from the Library of Congress.

ISBN 978-90-481-5422-7 ISBN 978-94-015-9566-7 (eBook)
DOI 10.1007/978-94-015-9566-7

Published by Kluwer Academic Publishers,
P.O. Box 17, 3300 AA Dordrecht, The Netherlands.

Sold and distributed in North, Central and South America
by Kluwer Academic Publishers,
101 Philip Drive, Norwell, MA 02061, U.S.A.

In all other countries, sold and distributed
by Kluwer Academic Publishers,
P.O. Box 322, 3300 AH Dordrecht, The Netherlands.

Printed on acid-free paper

Table of Contents

List of Figures

Preface

Quantum measurement (i.e., a measurement which is sufficiently precise for quantum effects to be essential) was always one of the most important points in quantum mechanics because it most evidently revealed the difference between quantum and classical physics. Now quantum measurement is again under active investigation, first of all because of the practical necessity of dealing with highly precise and complicated measurements.

The nature of quantum measurement has become understood much better during this new period of activity, the understanding being expressed by the concept of decoherence. This term means a physical process leading from a pure quantum state (wave function) of the system prior to the measurement to its state after the measurement which includes classical elements. More concretely, decoherence occurs as a result of the entanglement of the measured system with its environment and results in the loss of phase relations between components of the wave function of the measured system.

Decoherence is essentially nothing else than quantum measurement, but considered from the point of view of its physical mechanism and resolved in time. The present book is devoted to the two concepts of quantum measurement and decoherence and to their interrelation, especially in the context of continuous quantum measurement.

From the very first years of quantum mechanics quantum measurement has been permanently discussed as a problem and remains not completely solved even now. However, the character of the problem has radically changed. The well known 'measurement problem' in quantum mechanics was in fact metaphysical. Its solution was expected in proper wording, i.e., in the 'interpretation' of quantum mechanics. The situation gradually changed during last two decades. In highly precise measurements which have become feasible in practice, quantum effects are often essential, sometimes dominating. Therefore the most important goal is now a means of realizing quantum measurements, the correct description of realistic (and therefore complicated) measurements, the correct prediction of their results, and the behavior of the measured systems.

Thus quantum measurement has become a technical rather than a conceptual problem, and it is just in this aspect that it will be considered in the present book. The phenomena of entanglement and decoherence, very essential for the explanation of the physical nature of measurement, will be analyzed first. It will be shown how entanglement of the measured system with its environment leads to a partial loss of quantum features of the system, its decoherence. However, the main effort will be directed at the development of a simple but adequate mathematical description of quantum measurements and of the behavior of measured systems, including repeated or continuous measurements. Finally, continuous quantum measurements will be proved to be models of gradual decoherence. This closes the circle: a process leading to a quantum measurement is decoherence, but a continuous (repeated) quantum measurement may serve as a model for decoherence.

It has now been clearly understood that decoherence (i.e., partial loss of quantum properties) of a system may be induced not only by its interaction with an enormous number of degrees of freedom of a macroscopic measuring device, but also by an interaction with a few degrees of freedom which are external with respect to the system.

It must be noted in this connection that although decoherence by a macroscopic measuring apparatus or reservoir is usually quite rapid, it may be slow in the special conditions of almost isolated systems (small number of external degrees of freedom, a mesoscopic situation). Then details of the dynamics of decoherence become important. Continuous or repeated measurements arranged as a mesoscopic demonstration of the phenomenon of decoherence may serve as an example. One more example is a quantum computer, in which decoherence has to be avoided but cannot be avoided completely.

Having in mind situations of this type, we shall discuss in great detail the behavior of a system subject to gradual decoherence or continuous measurement (the dynamics of both these processes is in fact the same). If we are not interested in the physical nature of decoherence (measurement) but wish correctly to describe it, we need a phenomenological approach to decoherence allowing us to deal with decoherence without explicitly treating the environment of the measured system. We shall present the phenomenological approach to solving the problem. It is based upon the restricted path integral (RPI) or, equivalently, on the Schrödinger equation with a complex Hamiltonian (Men93). Other phenomenological approaches (such as stochastic Schrödinger equations) will be also reviewed.

Very interesting applications of the RPI approach which have been considered in the book (Men93) will be here omitted or mentioned only briefly, among them applications to quantum fields or relativistic particles. The reader interested in these applications as well as in general features of the whole approach must consult the book (Men93).

Instead we shall present here, and discuss in detail, new non-relativistic applications. The main one is the description of fuzzy continuous quantum measurements (FCQM) as opposed to the more widely known projective continuous measurements. The ability of FCQM to penetrate those quantum events which earlier were considered as non-separable will be illustrated by monitoring a quantum transition. It will be shown that the transition from one energy level to another under the influence, e.g., of a resonant laser pulse may be monitored (visualized) with the help of a FCQM.

It will be shown how such a measurement may be realized by a series of very weak short interactions. A sort of quantum analogue of the Central Limiting Theorem will be shown to take place in this situation: the behavior of a continuously measured two-level system is universal, independent of the details of the interactions. The universal behavior is presented by the simplest quadratic imaginary term in the effective Hamiltonian or by the Gaussian weight functional in RPI. This Gaussian evolution law and its generalizations give, in fact, a universal description of the dynamics of a system subject to continuous decoherence. The RPI analysis shows how this dynamics is governed by the information about the system recorded in its environment.

As has already been mentioned, conceptual problems connected with quantum measurements are not yet solved. We shall comment on them too, although very briefly because they, to our mind, lie outside physics and their solution is irrelevant for any properly physical problem.

Every effort was made to keep chapters, and even sections, of this book as independent of each other as possible. The reader is encouraged to skip the parts which are less interesting or are familiar to him and go directly to his main objective. When a return to some earlier material is necessary, cross-references will help the reader to do so in the most economic way.

The author acknowledges many fruitful discussions with his colleagues about the problems concerned in this book, among them H. von Borzeszkowski, V. B. Braginsky, S. Chumakov, S. Dürr, K.-E. Hellwig, E. Joos, V. A. Namiot, A. D. Panov, H. Rauch, G. Rempe and many others. It is a pleasure for me to recall that the program of the RPI approach to continuous measurements was started in the middle of 1970s under the influence of discussions with V. N. Rudenko, and the first paper was published at the recommendation of K. Thorne. Special thanks are due to A. Audretsch for his kind hospitality in Konstanz University where works on the continuous measurement of energy of a two-level system were completed.

Many long discussions with H.-D. Zeh were especially valuable in understanding the physical nature of decoherence, and his very deep paper of 1970 (Zeh70) became the key to the whole problem.

Yet it must be noticed, to avoid misunderstanding, that in the present book I have used (along with some other authors) the term 'decoherence' in a wider sense than H.-D. Zeh prefers to. He understands 'decoherence' as denoting uncontrollable and 'practically irreversible' decoherence caused by the interaction of the system with a macroscopically large number of degrees of freedom. This is assumed in Zeh's 'decoherence program' (GJK+96). However, no appropriate term exists for the loss of phase relations in a system influenced by a small number of external degrees of freedom (the mesoscopic situation). The quantum mechanical description of the phenomenon is the same both for a big environment of the system (a reservoir) and for small environment (a meter). Therefore usage of the term 'decoherence' in both these situations seems justified. Accordingly, the term 'environment' denotes all degrees of freedom which influence the system from outside, even if only a few of them are essential.

This terminology (widening of the words 'decoherence' and 'environment') is now accepted by many authors. It is quite natural and, in fact, unavoidable in the mesoscopic situation, which is quite important, particularly for applications. Nevertheless, it will be underlined several times in different connections in this book that 'decoherence' may mean either 'practically irreversible decoherence' induced by a macroscopic environment or 'reversible decoherence' induced by a microscopic environment.

Michael B. Mensky

Berlin, December 1999

Chapter 1

Introduction

Before starting a detailed exposition of the material we shall give in this chapter a sort of general survey of the book's content.[1] After sketching in Sect. 1.1 the plan of the book we shall then comment on some theoretical points which will be a subject of discussion in the book. First we shall give some idea of where the interrelated concepts of decoherence and quantum measurement appear and why they are important (Sect. 1.2). Then (Sect. 1.3) we shall briefly comment on the main aspects of the concepts to be considered: what is quantum measurement (decoherence); how fuzzy measurements may be described (in contrast to more familiar sharp measurements); and how continuous fuzzy quantum measurements may be represented, leading, in fact, to a simple model of gradual decoherence. At the end (Sect. 1.4) we shall concern ourselves with the most intriguing and most general view of decoherence as the appearance of classic features of the quantum world.

1.1. Motivation and the plan of the book

Continuous or repeated measurements of quantum systems have been actively discussed over the past decade — in the first place because this is where the intrinsic features of quantum theory are manifested to the utmost, and secondly because such measurements are steadily gaining practical importance (Dav76; Men83; BK92; Per93; Car93; Men93; GJK$^+$96; Kad97). Many years ago it was theoretically demonstrated (MS77; CSM77; Per80), and later confirmed experimentally (IHBW90) that often repeated measurements lead to freezing of the system in the original state (the so-called *quantum Zeno effect*).

If, however, the accuracy of each of repeated measurements is not high, then their effect on the measured system is not as strong, and the continuously measured system is not frozen (Men83; Men93). Recently it was demonstrated (AM97; AMN97) that a soft (fuzzy) continuous quantum measurement is capable of *monitoring a quantum transition.* The event of, say, transition between energy levels, which has usually been considered as

[1] In this we make use of the recent review (Men98b).

instantaneous in its occurrence, has turned out to be experimentally observable as unfolding in time. Concrete schemes of experiments were proposed in which the transition between two distinct quantum states can be monitored. Thus fuzzy (unsharp) continuous quantum measurements offer a new tool for the experimental study of quantum processes, capable of giving an insight deeper than previously deemed possible.

We shall discuss these new possibilities at the end of the book after presenting the theory of continuous quantum measurements. The first chapters will be devoted to the concept of quantum measurement generally. We will start in Chapter 2 with the discussion of the physical nature of quantum measurement. We are going to give a thorough analysis of the phenomenon of decoherence of a quantum system — the process in which a quantum measurement is realized. We shall see how and why the system acquires, in this process, classical features whilst its environment stores information about its state. The question of irreversibility of quantum measurements will be discussed in detail.

Special attention will be given in Chapter 3 to uncertainties induced in quantum measurements and their role in the so called Which Way experiments (showing how interference patterns characteristic of quantum mechanics may be lost as a result of some measurements). This type of experiment has been under intensive discussion during the last decade because of the suggestion that the concept of complementarity (wave–particle dualism) is more general than the uncertainty relation. It was claimed that in some cases the loss of an interference pattern after recording 'which way' information cannot be explained by uncertainty relations. We shall see that such a radical conclusion is not justified: even in the last very elegant Which Way experiment with the recording 'which way' information in the internal state of particles, the uncertainty relations appear and turn out to be sufficient for explaining the loss of the interference pattern (although the explanation which makes use of the concept of entanglement may in some cases be much more convenient).

Starting the analysis of continuous quantum measurements in Chapter 4, we shall demonstrate how the quantum Zeno effect arises in a series of sharp (projecting) measurements, and then discuss some models of unsharp continuous measurements, for example as in the process of quantum diffusion. We shall see that different mechanisms may lead to gradual decoherence of a quantum particle passing through a medium, but in all cases this process is equivalent to the continuous monitoring of the coordinate of the particle. The resulting equation is of the same type in all cases. This hints that the process of decoherence has a universal character.

The universality is very important because it makes possible to elaborate phenomenological approaches to the process of gradual decoherence or (what is in fact the same thing) of continuous measurement.

The phenomenological approaches to continuous quantum measurements will be presented in Chapter 5 and the subsequent chapters. Such approaches enable one to study continuous measurements without resorting to particular models of the environment (the measuring medium). We shall classify these approaches and discuss relations between them. Phenomenological approaches have the great advantage of being independent of the model and therefore universal. Besides, it turns out that a continuous measurement may itself serve as a model for the process of decoherence. Thus the phenomenological theory of continuous measurements provides in fact a *general phenomenological theory of decoherence.*

In great detail we shall discuss in Chapters 5, 6 the phenomenological approach based on restricted path integrals (quantum corridors), which in the special, but important, case of monitoring with absolute time resolution effectively reduces to the Schrödinger equation with a complex Hamiltonian. Subsequently in Chapter 7 we shall use this approach for analyzing continuous measurement of the energy of a multi-level system and — in greater detail — a two-level system. We shall prove that the fuzzy continuous measurement is capable of monitoring the quantum transition (Rabi oscillations), a process which is usually considered to be instantaneous. We shall see that the unfolding of the transition with time is experimentally observable, although, of course, with necessary quantum-mechanical restrictions on the reliability of this observation and with an unavoidable back influence of the measurement on the system.

In Chapter 8 it will be shown that a fuzzy continuous quantum measurement of a two-level system may be realized as a long series of short weak interactions of the system, and that a universal behavior of the continuously measured system (described by a simple quadratic imaginary potential) necessarily arises in this case, a sort of a quantum analogue of the Central Limiting Theorem of probability theory.

Chapter 9 presents a special, but very interesting, example of a continuous measurement of energy, a passive scheme of the quantum Zeno effect in the course of the Rabi transition onto an unstable level. The peculiarity of this setup is that it realizes a genuinely continuous rather than repeated measurement. Furthermore, this is an example of null (negative-result) measurements which have no classical analog.

From the conceptual point of view the restricted-path-integral approach is interesting in that it reveals an intriguing connection between information and dynamics: information recorded in the environment during a continuous measurement governs the dynamics of the measured system. This will be discussed in Chapter 10 together with other conceptual problems arising in connection with continuous quantum measurements.

More traditional conceptual problems of the quantum theory of measurements will be also commented upon in Chapter 10. The so called 'measurement problem' seems to be not really a physical, but rather a metaphysical problem. Nevertheless, we shall see that the restricted-path-integral approach, and the theory of imaginary potentials connected with it, sheds new light even on this problem, at least provide a more clear its formulation.

In the Appendix we shall consider some special questions of quantum mechanics which are important for the present book: density matrices; uncertainty relations; and path integrals.

1.2. Quantum measurements and decoherence

The relationship between quantum and classical descriptions of physical processes has been actively discussed since the early days of quantum mechanics, and still is. This issue is highly complicated because of the fundamental differences between the quantum and classical representations of physical systems and phenomena. The main distinction is apparently the quantum theoretical principle of superposition for the states of "corpuscular" systems, for which in classical physics superposition is not possible. Because of this, it is not quite true that classical theory gives an approximate description of something that is more precisely rendered in terms of quantum theory. Niels Bohr believed that a complete description of physical phenomena ought to include classical elements in addition to quantum representation.

Quantum mechanics assumes that, along with any two states of a quantum system $|\psi_1\rangle$, $|\psi_2\rangle$, there also exists their quantum mechanical (also known as coherent) superposition $|\psi\rangle = c_1|\psi_1\rangle + c_2|\psi_2\rangle$. If, however, the states $|\psi_1\rangle$, $|\psi_2\rangle$ differ considerably — being said to be classically distinct — for example, that they relate to the states of an elementary particle localized at points far from each other, then their superposition is not observed under ordinary circumstances. Such states may be referred to as non-coherent.

To be more precise, a superposition of macroscopically distinct states can be realized; but to have it survive for some time the system must be completely isolated. Even a very slight interaction with the environment will very quickly reduce the superposition $|\psi\rangle$ to one of the stable states $|\psi_1\rangle$ or $|\psi_2\rangle$, with only the respective probabilities $|c_1|^2$ and $|c_2|^2$ being known in advance. Such a conversion is the simplest example of *decoherence*. The fathers of quantum mechanics referred to this process as the reduction or collapse of the wave function. Today its nature is well understood, and now the dynamics of this process, its development in time is under investigation.

Two decades ago the process of decoherence was mostly of academic interest. Currently, however, the experimental techniques in the field, for example, of quantum optics (WM94) and neutron interferometry (RW00; Rau98) have become so much more refined that it has become possible to observe the formation of a superposition of macroscopically distinct states and the subsequent process of decoherence (Kni92; GK94; Zur91; HBR97). Moreover, the processes of decoherence must necessarily be included in the correct description of quantum systems interacting with the environment (open systems).

In particular, the processes of decoherence are important for the theory and practice of quantum computers (Unr95; Lan96). A quantum computer is a device capable of performing parallel computations by operating with a quantum superposition involving an enormous number of terms. In quantum computers (the elements of which have already been realized in practice) decoherence plays a dual role. In the course of quantum computations decoherence is an obstacle, since the superposition has to be prevented from falling apart. However, when the computation is complete, its result must be retrieved from the computer and represented in the classical (that is, the stable and steady) form. This is accomplished by the appropriate measurement of the state of the computer as a quantum system. In other words, the computer is made to interact with a special device which acts as a measuring system. This device measures certain parameters of the state of the computer, thus causing decoherence of this state. In this case the decoherence is brought about deliberately as a necessary final part of the whole quantum computation.

A hypothetical quantum computers is a device in which decoherence must play a crucial role. It is of no less importance, however, for many other quantum devices. The phenomenon of decoherence occurs whenever the system interacts or is made to interact with its environment, and the state of the system has some impact on the state of the environment. By observing the state of the environment one can then gain some information regarding the state of the system. Accordingly, interaction of the system with its environment may be interpreted as the measurement of the system. The information about the system is recorded in the environment. We then say that a measurement of the quantum system has taken place. The environment that performs this measurement can be created deliberately (measuring device or measuring medium), although in many cases it exists outside the experimenter's control, and often plays an undesirable role, leading to a special kind of dissipation.

It is important that such interaction with the environment inevitably modifies the state of the quantum system, causing decoherence. It turns out, however, that the behavior of the quantum system which is measured can be described with due account for its decoherence but without the need for explicit description of the measuring medium. The back effects of the environment are taken into account implicitly. This means that the system in question is treated as an *open system*, and its evolution is described *phenomenologically.*

The phenomenological theory of continuously measured quantum systems is an extension of the conventional quantum mechanics and considerably augments its capabilities. This theory is closed and intrinsically elegant.

In this way, the answers to the questions 'How does a quantum measurement occur?', and 'How does a continuously measured system behave?' bring us to the theory of open continuously measured quantum systems.

•

From a general course on quantum mechanics we know that the measurement of a quantum system obeys von Neumann's reduction postulate, which in the simplest case is represented by the scheme[2]

$$c_1|\psi_1\rangle + c_2|\psi_2\rangle \rightarrow \begin{cases} |\psi_1\rangle, & p_1 = |c_1|^2 \\ |\psi_2\rangle, & p_2 = |c_2|^2 \end{cases}$$

In this scheme $|\psi_1\rangle$ and $|\psi_2\rangle$ are the states each corresponding to a certain outcome of the measurement. According to the reduction postulate, one or the other measurement output are random events with respective probabilities p_1 and p_2, and the system assumes the corresponding state. This instantaneous transition, which cannot be described by the Schrödinger equation, is known as reduction or collapse of the state of the system. Mathematically, reduction can be described as the projection of the initial state vector either on the subspace of vectors proportional to $|\psi_1\rangle$ or on the subspace of vectors proportional to $|\psi_2\rangle$.

In the more general case the measurement is described by a set of projectors P_i, where the subscript i numbers the alternative outcomes of the measurement. If the measurement results in the ith alternative the initial state $|\psi\rangle$ after the measurement will go into the state $|\psi_i\rangle = P_i|\psi\rangle$ — that is, the reduction of a state is described by the corresponding projector. The probability that the measurement will result in the ith alternative is[3]

$$p_i = ||\psi_i||^2 = \langle\psi_i|\psi_i\rangle = \langle\psi_i|P_i|\psi\rangle. \tag{1.1}$$

[2]The term "quantum measurement" is not uniquely defined. In most of the present book it is assumed that a measurement supplies some information about a quantum system state in such a way that the measured observables actually lie in the limits specified by the measurement readout, not only before the measurement, but also during the measurement and after it.

[3]Usually the state after the measurement is described in terms of the normalized vector. We, however, prefer using the vector $|\psi_i\rangle$, since the reduction then is represented in a very straightforward manner, and the norm of the resulting vector is equal to the probability of the respective outcome of the measurement. This is especially convenient for describing repeated measurements.

This most simple scheme for describing measurement raises a number of questions, the first of which is how and why the collapse takes place. The answer to this question has been sought by many authors; see, for example, (Man50; Blo87; Kad97) and an excellent review of the topical literature in the book (GJK+96). Very clear analysis of the mechanism leading to collapse has been given by Zurek in (Zur81; Zur82).

This mechanism is now very well investigated, and the results of this investigation will be presented in the first part of the book. Briefly speaking, interaction of the measured system with the measuring apparatus (or, in a more general case, with the environment of this system) leads to quantum correlation (entanglement) between the states of the system and the apparatus (environment). Trying to describe only one part of this entangled state, the state of the measured system, we are forced to introduce a picture of the type of von Neumann projection. The system changes in such a way that a coherent superposition of some states ($|\psi_i\rangle$ from the above example) converts into their mixture. The process leading to loss of coherence was called 'decoherence'.

Perhaps the most interesting fact is that despite the radical clarification achieved during latter decades in the concept of quantum measurement, something mysterious is yet present in it. The point may be formulated in the following way: It is clear (and now seems almost trivial) how a coherent superposition is converted by the measurement into a mixture, but it cannot be explained in physical terms how selection is performed of a single term of this mixture.

We shall discuss this question (which is in a sense metaphysical) in the last part of the book (Chapter 10). However the main task will be consideration of actual physical problems: how the system evolves in the course of measurements of different types.

1.3. Continuous fuzzy quantum measurements

New questions arise when we consider a series of successive quantum measurements each of which is described by von Neumann's reduction. How does a system behave when subjected to a series of such 'instantaneous' (in fact, only very short) measurements? How does this behavior change if the intervals between the successive measurements tend to zero, so that the measurement becomes continuous? This book for the most part is devoted to the discussion of these questions.

The analysis of these problems reveals that repeated von Neumann measurements of an observable with a discrete spectrum (for example, measurement of the energy of a multi-level system) lead to the suppression of quantum transitions. As a result of continuous measurement the system is completely frozen at one point of the spectrum. This phenomenon has attracted much interest and became known as the *quantum Zeno effect* or the quantum Zeno paradox (MS77; CSM77; Per80).

Zeno paradox indicates that a continuous measurement may lead to triviality of the dynamics, in fact to its disappearance. But is that always the case? The answer is negative. In the first place, the system is not frozen if the measured observable has a continuous spectrum. Secondly, the dynamics remains non-trivial even in the case of a discrete spectrum, as long as the measurement is fuzzy (not too accurate). From this standpoint, a too accurate measurement of a quantum system is not advantageous, and this is a manifestation of the paradoxical nature of quantum mechanics.

The situation may be readily understood if we have in mind the picture of a repeated (sequential) measurement instead of a continuous one. The sequential measurement consists of a series of 'instantaneous' (in fact, very short) measurements. To avoid confusion, let us call a single measurement in the series an 'observation', keeping the word 'measurement' for the whole series of observations. Then a series of observations is a model of a continuous measurement. Then if each of the observations is precise enough, and they are often enough, the back influence of the whole series is strong. This may be called a hard (sharp, projecting) continuous measurement. Zeno paradox emerges only in a hard continuous measurement.

The reason for the Zeno type of behavior of the system is clear. If each observation in a series has high resolution, it can be well described by the von Neumann projection. Let the first of these observations project the system onto one of the eigenstates. If the period between two observations is small, the state changes insignificantly during this period, and most probably the system will be projected by the next observation onto the same eigenstate. Hence, the freezing of its evolution. The whole series of instantaneous (short) measurements then demonstrate the Zeno effect. This gives a picture of a hard continuous measurement. Sometimes such a series is called a 'projection continuous measurement'.

We shall be interested in another situation of a soft (unsharp, fuzzy) continuous measurement. If, as in the preceding argument, we represent the continuous measurement by a series of instantaneous observations, then each of the observations should have poor resolution (be fuzzy, or unsharp) and they should be not too often. Then the whole series of observations gives a model of fuzzy continuous measurement. Such a measurement is soft in the sense that its back influence on the measured system is not strong. The present book will be devoted to the theory of soft or fuzzy continuous quantum measurements.

It is evident from the preceding remarks that fuzzy measurements (among them continuous fuzzy measurements) have an advantage of rather weak back influence. The behavior of the measured system undergoing such a measurement is close to the behavior of the system in the absence of any measurement. This is why continuous fuzzy quantum measurements have turned out to be efficient instruments for monitoring quantum processes. A striking example is the monitoring of a level transition.

Transition of an atom between levels was conventionally considered as an instantaneous process (a quantum jump) even though the probability of the transition is a continuous function of time. The following arguments seemingly make unavoidable the interpretation of level transitions as quantum jumps.

If we perform frequently sharp observations of the system (as has been described above), with great probability we shall obtain each time the same result (the Zeno effect), the system will be repeatedly discovered to be at the same level. However, before each observation there is a small probability that the result of the observation will be different. This observation will then discover the system to be at another level. Subsequent observations will most likely discover the system at this new level, and this will again continue rather longer, until one of the observations gives a new result. Thus, we shall obtain the picture of the system staying at the same level for a rather long time, then an instantaneous transition to another level, staying for a long time at this new level, then a new 'quantum jump', and so on.

However, this argument is based on the assumption of precise (sharp) observations and quite short periods between them. With a fuzzy continuous measurement (consisting of unsharp observations with quite long periods between them) the picture will be quite different. Each observation will give only some vague information about the level the system is at. Actually the system will stay in the state of superposition of different levels, and each observation will slightly change the coefficients of the superposition.

With this picture of continuous measurement one can monitor the level transition without strong back influence on the system. This possibility and its experimental realizations will be considered in the present book.

There are different approaches to the description of open (continuously measured) quantum systems. Let us specify some of them:

- The model of measurement which includes the principal quantum system $\mathcal{S}$, its environment (or measuring device) $\mathcal{E}$, and the interaction between them.
- The equation for the density matrix (the master equation) of the system $\mathcal{S}$. It may be obtained from the model of the measurement after summation over degrees of freedom of the environment $\mathcal{E}$, but may also be derived independently of models.

– Restricted path integrals (quantum corridors), which can be reduced to the Schrödinger equation with a complex Hamiltonian.
– Stochastic Schrödinger equations.

The last two approaches (as well as the master equation if it is derived without any model) are phenomenological, they do not require a concrete model of the measurement. The abovementioned approach reducing a continuous measurement to repeated von Neumann's measurements is also phenomenological, but it is primitive and technically too complicated. Let us say a few words about two other phenomenological approaches.

One of them is based on a *master equation* for the density matrix ρ of the measured system. If the system has a Hamiltonian H and an observable A of the system is continuously measured (monitored), the density matrix ρ satisfies the equation

$$\dot{\rho} = -\frac{i}{\hbar}[H, \rho] - \frac{1}{2}\kappa[A, [A, \rho]]. \tag{1.2}$$

The reason why a density matrix arises instead of a wave function (a state vector) is evident. The continuously measured system is open because it continuously interacts with the measuring apparatus or measuring medium. Therefore, if we are interested only in the measured system we have to 'sum out' degrees of freedom of the measuring medium. The state of the rest degrees of freedom (of the measured system) cannot then be represented by a wave function, but only by a density matrix. The dynamics of the measured system is then described by Eq. (1.2).

It turns out, however, that the dynamics of the measured system may be described even by a wave function (a state vector) if we take into account the measurement readout (the so called selective approach). Let monitoring of the observable A give the function $a(t)$ as a readout. Then the state vector $|\psi\rangle$ of the measured system satisfies the following equation in which the Hamiltonian acquires, in fact, an imaginary additional term:

$$|\dot{\psi}\rangle = \left[-\frac{i}{\hbar}H - \kappa(A - a(t))^2\right]|\psi\rangle. \tag{1.3}$$

This equation may be derived from the approach based on restricted path integrals (Men79a; Men79b; Men93), elaborated by the author in accordance with the idea of Feynman (Fey48). However, it may also be derived in many other ways (as is the case for any fundamental equation). For example, it may be derived from the consideration of a concrete (but sufficiently general) scheme of realization of CFQM. We shall discuss all these questions in much detail.

With the help of Eq. (1.3) we shall investigate the continuous measurement of energy of a two-level system under the influence of the periodic driving field. It will be shown that the transitions between levels (Rabi oscillations) caused by this driving field may be monitored by the continuous measurement of energy provided this measurement is fuzzy enough. Such a phenomenological consideration has the advantage of simplicity. A concrete model will also be considered for a continuous measurement in a two-level system by means of a series of short interactions with a subsidiary system (a meter). This leads to the same conclusion as the phenomenological approach does.

Many other concrete measurement situations may be analysed in an analogous way. Some of them will be presented in this book. Some others, both nonrelativistic and relativistic, may be found in the previous book of the author (Men93).

Instead of Eq. (1.3) one may use an equivalent nonlinear stochastic equation. We shall briefly consider this approach too. In the stochastic equation there is no measurement output (such as the function $a(t)$ above). Instead, the influence of the measuring medium is represented by a white (Brownian) noise.

Finally, let us make two conceptual remarks.

The theory of open continuously measured quantum systems throws new light on the old question of whether or not quantum mechanics is a closed theory. The answer is affirmative if we are considering the Feynman formulation of quantum mechanics, which is extremely rich in ideas and includes the theory of restricted path integrals in it. This allows us not to introduce the theory of quantum measurements as a special independent postulate, but rather derive it from the Feynman formulation of quantum mechanics.

The phenomenology of continuously measured quantum systems leads to the conclusion about the dynamical role of information in the following sense. Of course, the dynamics of the measured system is determined by the nature of the measuring medium and its interaction with the system. All essential features of the dynamics, however, can be reconstructed if one knows only the information about the system which is recorded in the environment. It is information that determines the dynamics.

1.4. Emergence of classical and conceptual problems

We mentioned above (at the end of Sect. 1.2) a mysterious element still present in quantum measurements. It may be formulated as the mystery of *the emergence of the classical appearance of the quantum world.* This problem is widely discussed from different points of view. One of the best reviews, and further references, may be found in the book (GJK$^+$96).

The emergence of the classical picture includes the phenomenon of decoherence (the conversion of a coherent superposition to a mixture). Even in this phenomenon a very interesting aspect is present, namely irreversibility, strange enough in a theory based on the reversible Schrödinger equation. This aspect is completely understood now. Its origin is similar to the origin of irreversibility in the classical theory of multi-particle systems. Irreversibility of a quantum measurement is caused by entanglement of the measured system with an enormous number of degrees of freedom of its environment. Just as in the classical case, the process may, in principle, be reversed, but it cannot be reversed from any practical point of view.

The second, and really difficult for explanation, aspect in emergence of classical picture is *selection.* Even if decoherence converts a superposition into a mixture, one of the components of this mixture still has to be somehow selected to lead to a definite measurement readout. What might be a physical mechanism (if any) which could perform this selection? We shall consider this question in the last (shorter) part of the book. To our mind, the answer may be simpler than it might seem. In a sense, the creators of quantum mechanics were quite close to its solution. The theory of decoherence gives us a powerful instrument for the analysis of the problem, but the final conclusion is that the problem lies outside physics.

Simplifying the situation to its utmost, we may say the following. The measurement is associated with the selection of one out of many alternatives. We know well what happens when a particular alternative is selected, and can calculate the probability of each alternative. This gives answers to all questions that may arise in practice. How and why the selection is made, however, remains obscure. This question is hard to answer because this requires reconciling the quantum and the classical visions of the world. The attempts at solving this problem lead to very unusual constructions, the 'many worlds' interpretation of quantum mechanics being the main example (Eve73). This range of problems which may be regarded as the conceptual aspect of decoherence has been left out completely in most of the book, but will be briefly discussed at the end.

Part I

The Nature of Quantum Measurements

Chapter 2

The Physics behind Quantum Measurements

The simplest phenomenological description of an (instantaneous) measurement is based on von Neumann's projection postulate. Why does this very simple description turn out to be valid, how is it connected with physical processes in real setups realizing quantum measurements? Answers to these questions lead to the picture which is called 'decoherence'. The essential feature of decoherence is that phase relations between some components of the wave function are lost. We demonstrate this phenomenon with the help of a very simple model of measurement. It is demonstrated how interaction of the measured system with its environment may lead to its decoherence, i.e., partial loss of phase relations in the state of the system. A particular role of a reservoir (consisting of very many degrees of freedom) in the irreversibility of quantum measurements is discussed. However, we argue that decoherence may also occur in reversible situations. The latter will be illustrated in Chapter 3 by modern Which Way experiments.

Besides the simplest quantum measurement, we present a phenomenological description of more sophisticated measurement schemes, including fuzzy (unsharp) measurements by operators generalizing von Neumann's projectors. The connection between these operators and the resolution of the measurement is given in a very simple way. The difference between a selective and a non-selective description of quantum measurement is explained. Realization of fuzzy measurements is briefly discussed.

A very simple model of the process of decoherence is then given as a series of repeated fuzzy measurements. A more realistic re presentation of gradual decoherence as a continuous measurement will be given in Parts II and III of the book.

2.1. Measurement and decoherence

The behavior of a quantum system subject to measurement (idealized, of course) is described by von Neumann's reduction postulate (vN55). Let us consider very schematically the physical nature of the reduction (collapse) of the system's state. To illustrate the main idea it will suffice to analyse the simplest measurement amounting to the choice between two alternatives. Assume, for example, that an observable A which may take on one of the two values a', a'' is measured. Then, as a result of the measurement the system, with the appropriate probability, will go over, with the corresponding probabilities, into one of the eigenstates of the observable:

$$|\psi\rangle = c'|a'\rangle + c''|a''\rangle \to \begin{cases} |a'\rangle, & p_1 = |c'|^2 \\ |a''\rangle, & p_2 = |c''|^2 \end{cases} \tag{2.1}$$

This is the simplest form of von Neumann's reduction postulate, and the transition (2.1) from the state of the system before the measurement to its state after the measurement is the simplest form of the state reduction (wave function collapse). Later on (Sect. 2.3.1) we shall consider more complicated laws of reduction.

The same change of the state as in Eq. (2.1) may be expressed in terms of the density matrix (see Sect. A1) as the transition of the density matrix $\rho_0 = |\psi\rangle\langle\psi|$ (describing the pure state $|\psi\rangle$) into the density matrix of a mixed state:

$$\rho_0 = \begin{pmatrix} |c'|^2 & c'c''^* \\ c'^*c'' & |c''|^2 \end{pmatrix} \to \begin{pmatrix} |c'|^2 & 0 \\ 0 & |c''|^2 \end{pmatrix} = \rho. \tag{2.2}$$

The density matrix ρ contains the same information as the right hand side of (2.1), including information about the probabilities p_1, p_2 of the alternative measurement results and the corresponding final states $|a'\rangle$, $|a''\rangle$ of the system (see Sect. A1).

The physical process leading to the transition (2.2) is called *decoherence.* It results in the conversion of a superposition of a set of states $|a'\rangle$, $|a''\rangle$ to the mixture of the same states. The mark of decoherence is the disappearance of the nondiagonal elements of the density matrix:

$$\langle a'|\rho|a''\rangle = \langle a''|\rho|a''\rangle = 0. \tag{2.3}$$

Now let us discuss the physical mechanism which leads to decoherence. This mechanism consists in the interaction of the measured system with its *environment* (for example a measuring device) according to the scheme

$$\boxed{\textbf{System}} \leftrightarrow \boxed{\textbf{Environment}}$$

Interaction between the system and its environment leads to their entanglement, i.e., quantum correlation of the two systems (the measured system and its environment), so that the state of one of them contains information about the state of the other. As a consequence the measured system decoheres (is subject to decoherence). This is the way in which this process occurs.

As already mentioned, a situation equivalent to the measurement of a quantum system is often encountered even in the case where the measurement was not aimed by the experimenter but occurs as a result of some uncontrolled factors (say, the interaction with a reservoir). Therefore we shall speak of the environment $\mathcal{E}$ of our system $\mathcal{S}$ rather than of the device which measures the system.[1]

Assume that the environment prior to the interaction is in the state $|\Phi_0\rangle$. Interaction between $\mathcal{S}$ and $\mathcal{E}$ is referred to as measurement of $\mathcal{S}$ when it results in a specific correlation between $\mathcal{S}$ and $\mathcal{E}$, so that the information about the state of $\mathcal{S}$ is recorded in the state of $\mathcal{E}$. In our case the interaction between the two subsystems must translate the state $|a'\rangle\,|\Phi_0\rangle$ of the compound system into $|a'\rangle\,|\Phi'\rangle$, and the state $|a''\rangle\,|\Phi_0\rangle$ into $|a''\rangle\,|\Phi''\rangle$:

$$|a'\rangle\,|\Phi_0\rangle \to |a'\rangle\,|\Phi'\rangle, \qquad |a''\rangle\,|\Phi_0\rangle \to |a''\rangle\,|\Phi''\rangle. \tag{2.4}$$

Then the state of the environment after the measurement will tell us about the state of the measured system, at least if the measured system is in one of the states $|a'\rangle$ or $|a''\rangle$.

Let now prior to the measurement the composite system $\mathcal{S}+\mathcal{E}$ be in the state $|\Psi_0\rangle = |\psi\rangle\,|\Phi_0\rangle$, where $|\psi\rangle$ is a superposition of $|a'\rangle$ or $|a''\rangle$:

$$|\psi\rangle = c'|a'\rangle + c''|a''\rangle.$$

The result of the interaction between the two subsystems $\mathcal{S}$ and $\mathcal{E}$ is described (as is always the case in quantum mechanics) by a linear operator (a unitary evolution operator). Therefore, owing to the linearity of this operator and the correlation (2.4) the result of the interaction will be

$$\begin{aligned} |\Psi_0\rangle &= (c'|a'\rangle + c''|a''\rangle)\,|\Phi_0\rangle \\ &\to c'|a'\rangle\,|\Phi'\rangle + c''|a''\rangle\,|\Phi''\rangle = |\Psi\rangle. \end{aligned} \tag{2.5}$$

This is a so called *entangled state* of the compound system $\mathcal{S}+\mathcal{E}$. The subsystems $\mathcal{S}$ and $\mathcal{E}$ are entangled in this state. This means that the state of $\mathcal{S}+\mathcal{E}$ cannot be expressed as a product of a state of $\mathcal{S}$ with a state of $\mathcal{E}$. The entangled state also contains correlation between the states of $\mathcal{S}$ and $\mathcal{E}$ directly following from the simple and, in fact, classical correlation (2.4). However, the correlation in the entangled state is essentially quantum.

Many (if not all) specific features and paradoxes of quantum measurement are rooted in properties of entangled states. Einstein, Podolsky

[1]Notice that we apply the term 'environment' to the whole outside of our system. In the paper of Zurek (Zur82) the term 'environment' was introduced to denote only a subsystem having an enormous number of degrees of freedom and therefore providing irreversibility of the measurement. The microscopic subsystem directly interacting with the system was excluded from what was called 'environment'. We shall prefer the term 'reservoir' in this case. The role of a reservoir will be discussed in Sect. 2.2.1.

and Rosen were the first to analyse the paradoxical nature of such states (EPR35). Since then the Eistein–Podolsky–Rosen (EPR) paradox has bee discussed in many papers and books. The paradox itself does not now arise such interest because the nature of entangled states and their properties are well understood. Rather have quantum correlation of the EPR type or *EPR pair* become useful concepts for the analysis of quantum systems. In the analysis of quantum measurements they play the crucial role. In the present context the systems $\mathcal{S}$ and $\mathcal{E}$ form, in fact, an EPR pair.

Notice that, in contrast to the reduction of the states (2.1) or (2.2), the transition (2.5) occurs as a result of the conventional quantum mechanical evolution and can be described by the Schrödinger equation.

The next step is crucial in the chain of arguments leading to decoherence, yet it seems very simple and natural. We shall now go over to a description of the state in which only the system of interest $\mathcal{S}$ is included explicitly.

This is a very important (though seemingly trivial) moment in the quantum theory of measurement. It can be called *coarse graining*. We do not want to consider our system in all details, including its environment and the interaction with it. We need to abandon the details of the description connected with the environment. We apply the coarse grained description. However, the environment, even though excluded from the explicit description, of course influences our system, and this leads to decoherence of the system of interest as follows.

If we consider only the subsystem $\mathcal{S}$ without the explicit inclusion of its environment $\mathcal{E}$ in the consideration, we cannot do this in terms of state vectors only. Another instrument, the density matrix, is needed (see Sect. A1).

One can elaborate the density matrix ρ of a subsystem if the state $|\Psi\rangle$ of the composite system is known. The rule is to construct the density matrix $|\Psi\rangle\langle\Psi|$ of the composite system and calculate its trace with respect to all degrees of freedom except those of the subsystem of interest. The density matrix ρ of the subsystem is called (with respect to the density matrix of the total system) a *reduced density matrix.*

In our case we have to take the state $|\Psi\rangle$ of the system $\mathcal{S}+\mathcal{E}$ after the measurement, form the corresponding density matrix $|\Psi\rangle\langle\Psi|$, and calculate its trace with respect to the degrees of freedom of the environment $\mathcal{E}$. In this a way we shall obtain the density matrix of the measured system describing its state after the measurement. The result of the calculation is

$$\begin{aligned}\rho &= \mathrm{tr}_\Phi|\Psi\rangle\langle\Psi| \\ &= |c'|^2\,|a'\rangle\langle a'| + |c''|^2\,|a''\rangle\langle a''| \\ &+ \langle\Phi''|\Phi'\rangle\; c'c''^*|a'\rangle\langle a''| + \langle\Phi'|\Phi''\rangle\; c'^*c''|a''\rangle\langle a'|. \end{aligned} \tag{2.6}$$

In this expression the off-diagonal matrix elements are nonzero, so that we see no sign of decoherence. In fact, the environment of the measured system must have one more property in order it could cause decoherence (and therefore could realize a quantum measurement).

We have assumed already that the interaction between the system $\mathcal{S}$ and its environment $\mathcal{E}$ leads to the correlation (2.4). This is, however, not enough. It is necessary that the states $|\Phi'\rangle$, $|\Phi''\rangle$ of the environment which are correlated with the states $|a'\rangle$, $|a''\rangle$ of the measured system, be orthogonal:

$$\langle\Phi'|\Phi''\rangle = \langle\Phi''|\Phi'\rangle = 0. \tag{2.7}$$

If this is valid then the off-diagonal terms of the density matrix (2.6) vanish, and it takes the following form:

$$\rho = \mathrm{tr}_\Phi|\Psi\rangle\langle\Psi| = |c'|^2\,|a'\rangle\langle a'| + |c''|^2\,|a''\rangle\langle a''| \tag{2.8}$$

in accordance with Eq. (2.3). In this way the measurement leads to decoherence. The nature of this phenomenon has for quite some time been understood (see, for example, the excellent papers (Zeh70; Zeh73; JZ85)).

Let us say a few words about the condition of orthogonality (2.7). This condition is automatically satisfied in the situation of real measurement by a macroscopic device just because of the macroscopic character of this device. However this is not necessary for the orthogonality.

A large (macroscopic) number of degrees of freedom is a prerequisite for any measuring device, as is the fact that its states corresponding to different outcomes of the measurement (in our case $|\Phi'\rangle$ and $|\Phi''\rangle$) are '*macroscopically distinct*'. This means that the corresponding wave functions depend on very many variables and exhibit different functional dependence on the large number of these variables. The scalar product of such wave functions is practically equal to zero (to be more precise, it is exponentially small with an exponent of the order of -10^{23}). The reason is that the scalar product is an integral with respect to an enormous (macroscopic) number of variables. Even if the integral with respect to each separate variable is a little less than one, the total multiple integral will be close to zero. Hence, to a high degree of accuracy we have the property of orthogonality (2.7) satisfied.

However, the property of orthogonality may sometimes be provided even without a macroscopic number of degrees of freedom, in a microscopic environment. In this case decoherence of the system nevertheless occurs as a consequence of entanglement (quantum correlation). The difference occurring in this case is that the decoherence is still reversible (see Sect. 2.2.1).

We have not gone here into the details of the interaction between the measured system and its environment leading to the transition (2.5). Some examples of such interactions will be given later. The analysis of models of real measurements (see the book (GJK+96) and references therein) reveals that decoherence arises (that is, the off-diagonal terms vanish) exponentially fast in accordance with the expression

$$\left|\langle a''|\rho(t)|a'\rangle\right|^2 \sim e^{-\kappa(a''-a')^2 t}. \tag{2.9}$$

This occurs as more and more degrees of freedom of the environment become entangled with the measured system. As follows from Eq. (2.9), the characteristic *decoherence time* $t_{\rm d}$ is inversely proportional to the squared difference between the measured values of the observable,

$$t_{\rm d} \sim (a'' - a')^{-2}. \tag{2.10}$$

The same inverse quadratic dependence of the decoherence time on the difference between the measured values may be derived from the theory of continuous quantum measurements because a continuous measurement may be used as a model of decoherence (see Sect. 7.2.2 and, concretely, Eq. (7.16) for the 'level resolution time', a decoherence time in the concrete situation in which energy levels are measured.)

By the way, the dependence (2.10) explains why it is practically impossible to realize a superposition of two states of a particle which are localized at points far from each other. Even if such superposition were to arise, it would very quickly suffer decoherence owing to the interaction with the environment from which it cannot be completely isolated.

2.2. Irreversibility in reversible quantum mechanics

It is commonly believed that measurement in quantum mechanics is irreversible. In von Neumann's postulate and its generalizations measurement is described by an explicitly irreversible procedure, called reduction of the state or collapse of the wave function. This leads to an evident contradiction with the reversible character of the Schrödinger equation and the whole of quantum mechanics. This contradiction is, in fact, an origin of the so called 'measurement problem in quantum mechanics.

We have seen in Sect. 2.1 how the most important feature of quantum measurement, decoherence, results from the interaction of a system with its environment and the subsequent exclusion of the environment (i.e., the coarse grained description of the system). It can be seen from this, which will be further supported in Sect. 2.2.1, that irreversibility is not necessary for decoherence. Has measurement something in it besides decoherence? This is a difficult question and a root of the measurement problem. We shall briefly discuss this in Sect. 2.2.2 and then return to this rather philosophical question in Sect. 10.4 of Chapter 10 in the context of conceptual problems.

2.2.1. IRREVERSIBILITY: ENTANGLEMENT VS. MEASUREMENT

We saw in Sect. 2.1 that decoherence of the system $\mathcal{S}$ arises because of entanglement between the system and its environment. More precisely, the states of the system from the specified set of states (in the case considered these were $|a'\rangle$ and $|a''\rangle$) are correlated with the corresponding orthogonal states of the environment $\mathcal{E}$ (in the example considered the states $|\Phi'\rangle$ and $|\Phi''\rangle$). The specified states of the system are usually called 'pointer states'. They were orthogonal in the simple situation analysed in Sect. 2.1, but they are not necessary orthogonal, see Sect. 2.3. However, orthogonality of the corresponding states of the environment (i.e., the states of the pointer itself) is necessary. Just the orthogonality of $|\Phi'\rangle$ and $|\Phi'\rangle$ leads to decoherence between $|a'\rangle$ and $|a''\rangle$.

We mentioned in Sect. 2.1 that the orthogonality of alternative states of a macroscopic environment is always provided by the enormous number of its degrees of freedom (by its macroscopic nature).

Let us describe the structure of an environment in more detail. Distinguish a specific microscopic part of the environment directly interacting with the system and call it a *meter*. If we deal with the measurement deliberately set up, the meter is usually constructed in a special way to provide the necessary properties of the measurement. The meter in turn interacts with a macroscopic system containing an enormous number of degrees of freedom. Let us call this part of the environment a *reservoir*.[2] The scheme of the measurement is then the following:

Environment (over Meter and Reservoir)

System ↔ **Meter** ↔ **Reservoir**

If the measurement is organized deliberately, this scheme usually reflects well its organization and is advantageous for the analysis. In some other situations a subsystem playing the role of the meter cannot be defined in a natural way therefore the system interacts directly with the reservoir. This is the case for a particle moving through a medium, see Sect. 4.2.

We want to strongly underline that a (macroscopic) reservoir is not necessary for decoherence to occur. In fact, in the derivation of decoherence given in Sect. 2.1 the macroscopic character of the environment was not at all necessary. It is only orthogonality of the states $|\Phi'\rangle$ and $|\Phi''\rangle$ that was

[2] Zurek in the paper (Zur82) used the term 'environment' instead of our term 'reservoir', i.e., to denote only a subsystem with an enormous number of degrees of freedom providing the irreversibility of the measurement realized by a meter.

really employed. If it can be provided without the macroscopic nature of the environment, it is enough for decoherence to take place.

What, then, is the role of the (macroscopic) reservoir? It provides irreversibility or, better said, practical irreversibility.

To analyse this, return to the equation (2.5). It describes a unitary transformation leading to entanglement between the system and its environment. It is important here that the transformation is unitary. It results from the evolution described by the Schrödinger equation with some Hamiltonian, and it can be described by a unitary evolution operator. We can write

$$c'|a'\rangle\,|\Phi'\rangle + c''|a''\rangle\,|\Phi''\rangle = U(c'|a'\rangle + c''|a''\rangle)\,|\Phi_0\rangle \tag{2.11}$$

with some unitary evolution operator U. After such a unitary evolution we have an entangled state, and if, now, we are interested only in our system, we have to describe it by the reduced density matrix (see Sect. A1), and it demonstrates decoherence as a consequence of the orthogonality $\langle\Phi'|\Phi''\rangle = 0$.

However, any unitary evolution is reversible. In principle we may construct such conditions that the evolution of our compound system during some specified period of time will be described by the evolution operator $U' = U^{-1}$. After such an evolution the state (2.11) of the compound system will return to the disentangled (factorized) state which existed at the very beginning:

$$(c'|a'\rangle + c''|a''\rangle)\,|\Phi_0\rangle = U'(c'|a'\rangle\,|\Phi'\rangle + c''|a''\rangle\,|\Phi''\rangle). \tag{2.12}$$

Decoherence is now lost, and this is the result of a process which, in principle, can be realized.

We see, therefore, that in principle decoherence is always reversible. However, not in all cases it can be practically reversed. Decoherence is practically reversible when the environment is microscopic so that there is no reservoir, according to the following scheme:[3]

System $\leftrightarrow$ **Meter**

In this case we can actually construct conditions for reversing the process of decoherence. In the case of the Stern–Gerlach experiment, the beam of electrons, which has passed the zone of the magnet must be directed into the zone of one more magnet which should have an opposite sign of the field gradient and be identical otherwise. If between the two magnets there is no film or other macroscopic medium interacting with the electrons, the coherence will be recovered.

[3]Of course macroscopic parts of the environment are always present, but we mean that their influence is negligible.

In the general case we have a macroscopic reservoir (interacting with the meter or with the system itself in such a way that its influence cannot be neglected). In this case nothing changes in principle. The evolution (2.11) leading to entanglement is still unitary and can, in principle, be reversed. This means that the process leading to reversal could, in principle, exist, and it would not contradict any physical law including the laws of quantum mechanics. However, for constructing such a process one needs somehow to control a macroscopic number of degrees of freedom. This is, of course, impossible from the practical point of view. The situation is just the same as in the case of a broken cup in (classical) mechanics. The motion putting together all the pieces of the cup, and thus recovering the broken cup, is, in principle, possible, but it is in practice impossible to realize this motion.

Thus, the decoherence induced by entanglement with the microscopic environment (the meter) is practically reversible. If the decoherence is induced by entanglement with a macroscopic environment (the reservoir or the meter plus the reservoir), the decoherence may be considered reversible in principle, but it is irreversible from the practical point of view.

It may seem logical to use the word 'measurement' only in the second case, when the decoherence is practically irreversible, keeping the word 'decoherence' for a wider context when irreversibility is not necessary. Yet, this is a question of terminology. Besides, it is not evident that such a change would be convenient. Vice versa, it seems more convenient to extend the meaning of the word 'measurement' and use it each time when the information about the system state is transferred outside the system, resulting in decoherence of the system.[4] Indeed, it does not matter for the state of the system at the given time whether the information is kept in a small number of the outer degrees of freedom, and therefore decoherence is in practice reversible or it is distributed among a macroscopic number of degrees of freedom, making the recovery of coherence practically impossible.

Specifically, in Sect. 3.4 we shall deal with the situation when entanglement of a particle's spatial state with its internal state leads to the loss of the interference pattern. The process may be reversed until we have observed the loss of the interference pattern; however, all the features of the decoherence are just the same as they would be in the case of irreversible measurement.

[4]To be more precise, decoherence occurs if the information about the state of a number of degrees of freedom is recorded in the state of some other degrees of freedom. In reality both sets of degrees of freedom may belong to the same physical body. The simplest example is the Stern-Gerlach experiment in which spin degree of freedom of a particle is considered as a measured system and its center of mass as the environment. One more interesting example will be considered in Sect. 3.4.

2.2.2. THE MANY WORLDS INTERPRETATION OF QUANTUM MECHANICS

We have seen that irreversibility is not a principle of quantum measurement but rather a convenient approximation for its description. There is a quite original attempt to overcome irreversibility completely, the so called 'many worlds' interpretation of quantum mechanics proposed by Everett (Eve73). It may be explained with the help of the simple two-alternatives measurement analysed in Sect. 2.1.

Starting from the state state

$$|\Psi_0\rangle = (\alpha|a\rangle + \beta|b\rangle)|\Phi_0\rangle$$

of the system $\mathcal{S} + \mathcal{E}$ and applying only quantum mechanics with its Schrödinger equation and unitary evolution, we can arrive at the state

$$|\Psi\rangle = \alpha|a\rangle|\Phi_a\rangle + \beta|b\rangle|\Phi_b\rangle = \alpha|\mathcal{A}\rangle + \beta|\mathcal{B}\rangle. \tag{2.13}$$

If now we ignore the state of the environment $\mathcal{E}$, we can obtain a coarse grained description of the system $\mathcal{S}$ including the picture of its decoherence. In this picture both alternatives are present stochastically, each of them is characterized by the corresponding probability. This may be described in a non-selective way, with the help of the density matrix, or in a selective way, where each of the alternatives is presented separately from each other. However, there is no justification or explanation of the mechanism of selecting one of the alternatives. However, we know that it is only one of the alternatives, not all of them, that is observed in any real experiment. This is the main conceptual difficulty of quantum mechanics leading to what is called 'measurement problem'.

From practical point of view it is not necessary to include in the theory any explicit description of a mechanism leading to the selection one of the alternatives. All that we need for real calculations is probabilities of alternatives. However, from a purely theoretical (or rather philosophical, or metaphysical) point of view it may be inquired how selection of one of the alternatives is performed, and what is the mechanism for doing so. Quantum mechanics has no mechanism inside itself.

Being logical and yet remaining within quantum mechanics, one is forced to conclude that *no selection at all is performed* and that both alternatives exist (or many alternatives in the general case). The question is then to be answered of why an observer sees only one alternative in practice. The answer given by Everett is that only one of the alternatives describes our world. The second alternative (any other alternative in the general case) describes a possible world too. No one of the worlds is better than another in any respect. In each of them there are their own observers, but their observations differ.

This strange picture of parallel worlds is logically quite possible. Whether it solves the problem is for the reader to decide. There is an opinion that it solves nothing. One has always to ask whether a theory can, in principle, be falsified, that is to say, whether it is testable. Referring to Everett's many worlds interpretation it seems to be impossible to verify the existence of many worlds (corresponding to the many alternatives of a quantum measurement) by methods used in science. In this sense the question does not lie within the scope of physics. We shall now try to justify this statement. Afterwards we shall return to the question when discussing conceptual problems in Sect. 10.4 of Chapter 10.

To deal with the clear formulation of the problem let us ask how can one find out, with the help of a measurement, whether two alternatives are present in a superposition or only one of them. In other words, we need a measurement which can discover whether both coefficients α, β are nonzero in Eq. (2.13).[5]

An observer acting in the usual way will discover only one of these alternatives. For example, he could perform a measurement distinguishing between the states $|\mathcal{A}\rangle$, $|\mathcal{B}\rangle$ (in terms of Sect. 2.3, the measurement of the pair of projectors $P = |\mathcal{A}\rangle\langle\mathcal{A}|$, $Q = |\mathcal{B}\rangle\langle\mathcal{B}|$). He will then obtain one of the two results with the corresponding probabilities $|\alpha|^2$ and $|\beta|^2$. After a single measurement of this type the observer does not know whether the second alternative existed or not. Of course, repetition of the same measurement would finally give approximate estimates for $|\alpha|^2$ and $|\beta|^2$, thus answering the question about the existence of both alternatives. However, even this repeated measurement does not distinguish between a real coherent superposition and mixture of the same states.

Indeed, if the state subject to the measurement $\{P, Q\}$-is a mixed state represented by the density matrix

$$\begin{aligned} \mathcal{R} &= |\alpha|^2|\mathcal{A}\rangle\langle\mathcal{A}| + |\beta|^2|\mathcal{B}\rangle\langle\mathcal{B}| \\ &= |\alpha|^2 P + |\beta|^2 Q, \end{aligned} \tag{2.14}$$

the statistics of a series of the repeated measurements (with the same state (2.14) prepared before each measurement) will be just the same as for the superposition (2.13). This follows both from the formal definitions of Sect. 2.3 and from the interpretation of the density matrix (2.14): the system is in one of the states $|\mathcal{A}\rangle$, $|\mathcal{B}\rangle$, but it is not known which one of them. Instead the corresponding probabilities $|\alpha|^2$, $|\beta|^2$ are given (Sect. A1).

[5]This is not precisely what we need when asking about the existence of Everett's many worlds, but this is the only relevant question which can be explored by the methods of physics.

Our aim is just to distinguish between a coherent superposition and decoherent mixture, therefore the measurement $\{P, Q\}$ does not solve the problem. To solve it, we need an operator including both alternatives simultaneously, for example, one of the operators

$$S_1 = |\mathcal{A}\rangle\langle\mathcal{B}|, \quad S_2 = |\mathcal{B}\rangle\langle\mathcal{A}|.$$

Their mean values

$$\langle\Psi|S_1|\Psi\rangle = \alpha^*\beta, \quad \langle\Psi|S_2|\Psi\rangle = \alpha\beta^*$$

are equal to zero only if one of the alternatives is actually absent in the superposition. The mean values of these operators in the mixed state (2.14) is zero:

$$\langle S_i\rangle_{\mathcal{R}} = \mathrm{tr}(\mathcal{R}\, S_i) = 0.$$

Thus, mean values of these operators distinguish between a coherent superposition and a mixture.

The operators S_1, S_2 are non-Hermitian but they can be expressed through the two Hermitian operators as $(H_1 \pm iH_2)/2$, and the latter operators

$$\begin{aligned} H_1 &= |\mathcal{A}\rangle\langle\mathcal{B}| + |\mathcal{B}\rangle\langle\mathcal{A}|, \\ H_2 &= -i\,(|\mathcal{A}\rangle\langle\mathcal{B}| - |\mathcal{B}\rangle\langle\mathcal{A}|) \end{aligned}$$

can be measured to estimate $\alpha^*\beta$ and find out whether both components of the superposition are present or not. It is not a single measurement but a long series of measurements on the same state $|\Psi\rangle$ that may give us the mean value (2.2.2). We have to prepare the same superposition each time in order to be able to investigate it repeatedly.

If the environment $\mathcal{E}$ is simple (i.e., only a meter, in terms of Sect. 2.2.1), then the structure of the states $|\mathcal{A}\rangle$, $|\mathcal{B}\rangle$ is quite simple so that the measurement of the operator $|\mathcal{A}\rangle\langle\mathcal{B}|$ can be arranged. This (or something similar) is what is done in the experiments involving 'Schrödinger cats' (Kni92; GK94). In such experiments superpositions of the states are prepared which in ordinary conditions rapidly decay into a mixture (decohere). Typically this may be a superposition of two coherent states of an oscillator. If the system is strictly isolated, then superpositions live during quite a long time without decoherence and may be explored during this period.

If, however, the superposition (2.13) arises in the interaction of the system with a complicated environment (including a reservoir, in the terminology of Sect. 2.2.1), then the structure of the states $|\mathcal{A}\rangle$, $|\mathcal{B}\rangle$ may be presented as follows (GJK+96):

$$\begin{aligned} |\mathcal{A}\rangle &= |a\rangle|A_1\rangle|A_2\rangle\ldots|A_m\rangle\ldots|A_M\rangle, \\ |\mathcal{B}\rangle &= |b\rangle|B_1\rangle|B_2\rangle\ldots|B_n\rangle\ldots|B_N\rangle \end{aligned} \tag{2.15}$$

with a macroscopically large M and N (of the order of 10^{23}). In this case it is practically impossible to organize the measurement of the operator $|\mathcal{A}\rangle\langle\mathcal{B}|$.

Any mistake in the measured operator will make the measurement meaningless. If we omit even a single factor present in Eq. (2.15) when arranging the measurement, the measured operator will be $S_1' = |\mathcal{A}'\rangle\langle\mathcal{B}|$ instead of S_1, where $|\mathcal{A}'\rangle$ differs from $|\mathcal{A}\rangle$ in that one factor, say $|A_m\rangle$, is absent. It is easily verified that the mean value of such an operator is automatically zero,

$$\langle\Psi|S_1'|\Psi\rangle = \langle\Psi|\mathcal{A}'\rangle\langle\mathcal{B}|\Psi\rangle = 0$$

The measurement of S_1' will appear as if there were no superposition even if there were a superposition in reality. The same is valid of course if $S_1' = |\mathcal{A}\rangle\langle\mathcal{B}'|$ with $|\mathcal{B}'\rangle$ differing from $|\mathcal{B}\rangle$ by at least a single factor $|B_n\rangle$.

It is remarkable that the usual measurements distinguishing between the alternatives do not have this type of obstacles. Measuring the pair of projectors $|A_m\rangle\langle A_m|$, $|B_n\rangle\langle B_n|$ with arbitrary m, n gives just the same result as measuring of the whole $|\mathcal{A}\rangle\langle\mathcal{A}|$, $|\mathcal{B}\rangle\langle\mathcal{B}|$.

The conclusion is that the identification of one of the alternatives is possible with the help of an arbitrary consequence of this alternative (for example with the help of a photon radiated by a measuring device and then sensed by an observer's eye). In contrast, the confirmation of the fact that both alternatives are present in a coherent superposition (after the interaction between the system and its environment) is possible only after many repeated experiments (with the original situation each time restored) under the necessary condition that *the whole chain of consequences of the alternatives is controlled.*

This is practically impossible in the case of those superpositions which arise in an interaction with a reservoir (in practice irreversible measurements, see Sect. 2.2). Whether the corresponding alternatives exist in a coherent superposition or in a mixture cannot be distinguished experimentally. One must therefore say that the existence of the Everett's parallel worlds is not testable.

2.3. Phenomenology and realization of measurements

From the simplest example of a quantum measurement with two alternatives let us go over to a more realistic type of measurements, finishing with the case of a fuzzy measurement with continuous alternatives. The constructions are quite natural, but may seem abstract. However, we shall see that even the most general of them, the description of a fuzzy measurement with a continuous set of alternatives, may be realized by a quite simple Hamiltonian.

2.3.1. VON NEUMANN PROJECTORS AND THEIR GENERALIZATIONS

Two alternatives

The measurement considered in Sect. 2.1 distinguishes between two alternatives according to the scheme

$$|\psi\rangle = c_1|a_1\rangle + c_2|a_2\rangle \rightarrow \left\{ \begin{array}{ll} |a_1\rangle, & p_1 = |c_1|^2 \\ |a_2\rangle, & p_2 = |c_2|^2 \end{array} \right. \tag{2.16}$$

The two alternatives may be presented by the following two projectors (in the two-dimensional state space spanned by the vectors $|a_1\rangle$, $|a_2\rangle$)

$$\{P_1, P_2\}: \quad \begin{array}{ll} P_1|a_1\rangle = |a_1\rangle, & P_1|a_2\rangle = 0 \\ P_2|a_1\rangle = 0, & P_2|a_2\rangle = |a_2\rangle \end{array}$$

Then the scheme (2.16) of the measurement may be conveniently rewritten in the form

$$|\psi\rangle \rightarrow \left\{ \begin{array}{ll} |\psi_1\rangle = P_1|\psi\rangle, & p_1 = ||\psi_1||^2 = \langle\psi|P_1|\psi\rangle \\ |\psi_2\rangle = P_2|\psi\rangle, & p_2 = ||\psi_2||^2 = \langle\psi|P_2|\psi\rangle \end{array} \right. \tag{2.17}$$

Many alternatives

The generalization to the case of arbitrary number of alternatives is now evident. The alternatives are represented by an arbitrary complete set of orthogonal projectors (in the state space)

$$\{P_i\}, \quad P_i^2 = P_i, \quad P_iP_j = 0,\ i \neq j, \quad \sum_i P_i = \mathbf{1} \tag{2.18}$$

A change of a state under the measurement is then described by the scheme

$$\begin{array}{rcl} |\psi\rangle & \stackrel{i}{\rightarrow} & |\psi_i\rangle = P_i|\psi\rangle, \\ p_i & = & ||\psi_i||^2 = \langle\psi|P_i|\psi\rangle \end{array} \tag{2.19}$$

Notice that here we make use of unnormalized vectors $|\psi_i\rangle$ to represent the final states. This gives great technical advantages. In particular, such a vector has a simpler form than the corresponding normalized vector, and it contains information not only about the final state but also about the probability of the realization of this state. The simplification seems not so important, but it becomes very important when a long chain of measurements performed one after another is considered.

Hints from measuring the coordinate

We shall now consider the most realistic class of so called *fuzzy measurements*, which will be discussed in much detail in the rest of the book. The two further steps in the generalization of von Neumann's projectors necessary for our aim follow from the problem of measuring the coordinate (although measuring any other observable with a continuous spectrum has the same features).

Measurement of the coordinate with a finite resolution Δa can be fitted into the preceding scheme with the help of the following trick (Fig. 2.1, left hand part). Let us divide the q-axis into equal intervals of length

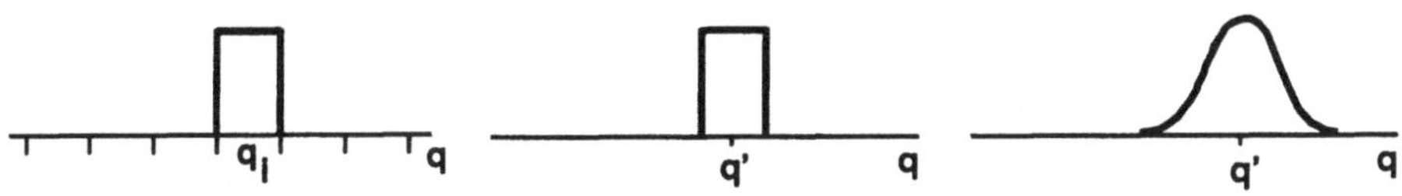

Figure 2.1. Measurement of the coordinate: discretization (left), overlapped projectors (middle), and overlapped smooth characteristics (right).

$2\Delta a$. Let q_i be the centers of these intervals. Then we can assume that the measurement of the coordinate may give only one of the results (thus introducing discrete alternatives) and put in correspondence with these results the projectors P_i defined as follows:

$$(P_i\,\psi)(q) = \begin{cases} \psi(q), & \text{if} \quad |q - q_i| < \Delta a, \\ 0 & \text{otherwise.} \end{cases}$$

The projector P_i simply multiplies the wave function $\psi(q)$ by the characteristic function of the interval $[q_i - \Delta a, q_i + \Delta a]$ (that is the function equal to unity on the points of the interval and zero outside). This function serves as a characteristic function of the measurement.

The measurement of the coordinate with resolution Δa is then described by the scheme (2.19), but this description is unsatisfactory in the following two respects: 1) the measurement's results are discrete; and 2) the measurement's characteristic is rectangular. Of course, we expect that: 1) the measurement of the coordinate may give an arbitrary real number q' as a result; and 2) such a result points out that the actual coordinate q lies somewhere close to q' with a probability smoothly decreasing with growth of $|q - q'|$. This means that the set of alternatives must be continuous, separate alternatives being enumerated by the continuous parameter q', and that the transformation of the system's state after the measurement with this result is described by multiplication by some smooth curve:

$$(R_{q'}\,\psi)(q) = R_{q'}(q)\psi(q).$$

According to these arguments we can further generalize the scheme of measurement. We shall do this in two steps.

Smooth alternatives

Consider, first, discrete alternatives but with smoothed characteristics. We shall present the set of alternatives by operators $\{R_i\}$ not requiring them to be orthogonal projectors. Operators $\{R_i\}$ may be arbitrary but the sum of their 'square moduli' must be equal to the unit operator (for correct normalization):

$$\{R_i\}, \quad \sum_i R_i^\dagger R_i = 1 \tag{2.20}$$

The measurement is now described by the scheme

$$\begin{aligned} |\psi\rangle &\xrightarrow{i} |\psi_i\rangle = R_i|\psi\rangle, \\ p_i &= ||\psi_i||^2 = \langle\psi|R_i^\dagger R_i|\psi\rangle. \end{aligned} \tag{2.21}$$

The correct normalization

$$\sum_i p_i = 1$$

is provided by the condition (2.20). Products of different operators from the set $\{R_i\}$ is now not necessarily zero. This means that different measurement results may lead to overlapping states. Such a measurement is called *fuzzy* or *unsharp*. This is a realistic property, especially for measuring observables which have continuous spectra (such as a coordinate).

Continuous set of smooth alternatives

One more step leads to the description of the most realistic type of measurements, a fuzzy measurement with a continuous set of alternatives. Alternatives are enumerated now by a continuous parameter a' and presented by a continuous set of operators. The sum is, of course, replaced by an

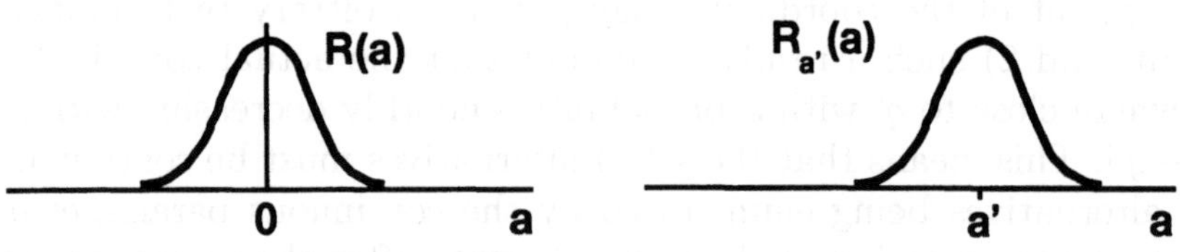

Figure 2.2. Measurement of an arbitrary observable A: the characteristic function of the measurement (left), and the characteristic function for the given measurement result (right).

integration,[6]

$$\{R_{a'}\}, \quad \int da' R^{\dagger}_{a'} R_{a'} = \mathbf{1}. \tag{2.22}$$

According to Eq. (2.22) the set of positive operators $E(a') = R^{\dagger}_{a'} R_{a'}$ form a decomposition of the unit operator. This is analogous to the definition of a measure on some set with the help of a weight function defined on that set,

$$\int f(a')\, d\mu(a') = \int f(a')\, w(a')\, da'.$$

The weight function then forms a decomposition of unity,

$$\int w(a')\, da' = 1.$$

This is why the mapping $a' \to E(a')$ is called in the literature a positive-operator-valued (POV) measure on the set $\{a'\}$ of all possible values of the parameter a'.

The measurement is now represented by the scheme

$$\begin{aligned} |\psi\rangle \quad &\xrightarrow{a'} \quad |\psi_{a'}\rangle = R_{a'}|\psi\rangle, \\ p_{a'} \quad &= \quad ||\psi_{a'}||^2 = \langle\psi|R^{\dagger}_{a'} R_{a'}|\psi\rangle. \end{aligned} \tag{2.23}$$

As usual with a continuous spectrum, the entity $p_{a'}$ is here not a probability but a probability density. To obtain the probability that the parameter a' turns out to belong to the set $\mathcal{A}$ one has to integrate over this set:

$$\text{Prob}[a' \in \mathcal{A}] = \int_{\mathcal{A}} da'\, p_{a'}. \tag{2.24}$$

We shall see below (2.32) how the resulting state may be described if we know that the result of the measurement is in the set $\mathcal{A}$.

Let us make the formula (2.23) more concrete. For simplicity we shall use the A-representation of the wave function, so that the state $|\psi\rangle$ is represented by the wave function $\psi(a)$. The operator $R_{a'}$ describing the measurement giving the result a' will be taken in the form

$$R_{a'}\psi(a) = R(a - a')\psi(a), \tag{2.25}$$

where $R(a)$ is a function concentrated in some vicinity of zero and representing the 'fuzziness' of the measurement. It is a *characteristic function*

[6]A number of parameters may, of course, be instead of a single parameter a', but the generalization to this case is straightforward. It may sometimes be convenient to integrate over the parameters with some weight.

of the measurement and may, for example, be (and typically is) a Gaussian function.[7] The width of this function corresponds to the resolution of the measurement. Normalization (2.22) of the operators $R_{a'}$ is provided by the corresponding normalization of the function $R(a)$ as follows:

$$\int da\, |R|^2(a) = 1. \tag{2.26}$$

In many important examples the operators $\{R_{a'}\}$ (or operators $\{R_i\}$ above) are Hermitian and positive.[8] They correspond to a non-negative function $R(a)$. The measurements described by such operators are minimally disturbing. This means that the measurement changes an initial state scaling the modulus of the wave function (by a factor depending on the argument of the wave function) but not changing its phase.

An example of a fuzzy measurement of a coordinate may be illuminating. If the measurement is described by the Gaussian function

$$R(q) = \mathcal{N}\, e^{-\frac{q^2}{4\lambda^2}},$$

it is a minimally disturbing measurement of the coordinate with resolution λ. If, however, the characteristic function of the measurement is

$$R(q) = \mathcal{N}\, e^{-\frac{q^2}{4\lambda^2}+\frac{i}{\hbar}pq},$$

it is a coordinate measurement of the same resolution but accompanied by the transition of the linear momentum p.

Fuzzy measurement of a discrete observable

We have derived the idea of fuzzy measurements in trying to describe how to measure observables with continuous spectrum. Nevertheless, the resulting definition is applicable to observables with discrete spectra too. If A is an observable with the spectrum $\{a_i\}$ then the role of the wave function in the A-representation is plaied by the coefficients c_i of the expansion

$$|\psi\rangle = \sum_i c_i |a_i\rangle.$$

Therefore, the operators $R_{a'}$ may be defined as follows:

$$R_{a'}\, |\psi\rangle = \sum_i R(a_i - a')\, c_i |a_i\rangle \tag{2.27}$$

[7]The function $R_{a'}(a)$ may be in the r.h.s. of Eq. (2.25) instead of $R(a - a')$. There is then no general characteristic function, but it is defined separately for each result of the measurement. However, a general characteristic function as in Eq. (2.25) is the most typical.

[8]A Hermitian operator R is called positive ($R \geq 0$) if its mean value in an arbitrary state $|\varphi\rangle$ is non-negative, $\langle\varphi|R|\varphi\rangle \geq 0$.

where $R(a)$ is a characteristic function of the measurement. It is interesting that the measurement may have a continuous set of results, even for a discrete spectrum of the measured observable. This is, of course, a consequence of fuzziness of the measurement. We shall see in Sect. 2.4 that a series of such measurements leads finally to a sharp (von Neumann type) measurement of A, and the sequence itself forms a model of decoherence leading to the von Neumann's reduction.

Non-selective description

In the above formulas we have made use of a selective description of different types of measurements. This means that the final state after the measurement was pointed out for a certain result of the measurement. This is why the final state of the system has always been pure, provided that its initial state was pure.

The same measurement may be described in a non-selective way when we do not know which of the alternative results of the measurement is the actual one. In this case all possible results of the measurement are taken into account in the description of the final state.

The non-selective description may be obtained if we rewrite the selective description in terms of density matrices (see Sect. A1) and then sum over all alternatives. For discrete alternatives we then have instead of Eq. (2.21)

$$|\psi\rangle\langle\psi| \xrightarrow{\{i\}} \sum_i R_i^\dagger |\psi\rangle\langle\psi| R_i. \tag{2.28}$$

It can be easily seen that the contents of this formula is just the same as in Eq. (2.21).

In particular, this formula again starts with a pure state (although represented by a density matrix), but the final state is mixed. The reason for this is our incomplete knowledge. In principle we could know the result of the measurement because it is recorded (in the environment) and can be read off, but we do not want to gather this information or cannot do so owing to some practical reasons (not because of some prohibition). Therefore our knowledge is incomplete, and the state is described by a density matrix, not a vector.

Now, however, we can easily deal with the case when the initial state is also mixed. Let it be described by a density matrix

$$\rho_0 = \sum_c q_c |\varphi_c\rangle\langle\varphi_c|. \tag{2.29}$$

This means that the system is in one of the pure states $|\varphi_c\rangle$ with the corresponding probabilities q_c. Applying Eq. (2.28) to each of these pure states and using the usual rules of operations with probabilities we have the following general formula:

$$\rho_0 \xrightarrow{\{i\}} \sum_i R_i^\dagger \rho_0 R_i. \tag{2.30}$$

It is evident how one can generalize this formula to the case of the continuous set of alternatives (2.22). The description of the measurement will differ from Eq. (2.30) only in integration instead of summation:

$$\rho_0 \xrightarrow{\{a'\}} \int da' R_{a'}^\dagger \rho_0 R_{a'}. \tag{2.31}$$

An intermediate situation arises when we know that the result of the measurement a' belongs to some set $\mathcal{A}$. In this case the description of the measurement is partly non-selective. The state of the measured system must be described by a density matrix given by an integral of the type of (2.31) but with integration over the set $\mathcal{A}$:

$$\rho_0 \xrightarrow{\mathcal{A}} \rho = \int_{\mathcal{A}} da' R_{a'}^\dagger \rho_0 R_{a'}. \tag{2.32}$$

This intermediate situation is, of course, the most typical. The probability that the result of the measurement belongs to the set is given by Eq. (2.24). If the set $\mathcal{A}$ is sufficiently narrow then the operators $R_{a'}$ contributing to Eq. (2.32) differ from each other only insignificantly. In this situation the final (after the measurement) state ρ in Eq. (2.32) is close to the pure state $R_{a'}|\psi\rangle$ provided that the initial state ρ_0 was pure $|\psi\rangle$. The scheme (2.23) may then be considered not only as an intermediate step before the final formula (2.32) but also as a good approximation of an actual transformation of the state in the measurement.

2.3.2. REALIZATION OF PROJECTING MEASUREMENTS

We have seen already in Sect. 2.1 how the simplest quantum measurement selecting between two states $|a_1\rangle$ and $|a_2\rangle$ may be arranged. With many alternative vectors $|a_i\rangle$, we have practically the same arangement. Starting from the initial state $|\Phi_0\rangle$ the environment $\mathcal{E}$ must be transferred (by the interaction with the measured system $\mathcal{S}$) into a correlated state $|\Phi_i\rangle$. Therefore the evolution (governed by the Hamiltonians of the systems $\mathcal{S}$, $\mathcal{E}$ and their interaction Hamiltonian) must result (during a specified period) in the transformation

$$|a_i\rangle\,|\Phi_0\rangle \xrightarrow{U} |a_i\rangle\,|\Phi_i\rangle. \tag{2.33}$$

Then owing to the linearity of the evolution operator U, we have for the same initial state of the environment but for an arbitrary state of the system,

$$|\psi\rangle = \sum_i c_i |a_i\rangle.$$

the following evolution law:

$$|\psi\rangle\,|\Phi_0\rangle = \sum_i c_i |a_i\rangle\,|\Phi_0\rangle \xrightarrow{U} \sum_i c_i |a_i\rangle\,|\Phi_i\rangle = |\Psi\rangle.$$

Taking the trace over the degrees of freedom of the environment and making use of the orthogonality of the states $|\Phi_i\rangle$,

$$\langle\Phi_j|\Phi_i\rangle = \delta_{ji}, \tag{2.34}$$

we have the following reduced density matrix (see the Appendix A1) for the final state of the system $\mathcal{S}$:

$$\begin{aligned} \rho &= \mathrm{tr}_\Phi |\Psi\rangle\langle\Psi| \\ &= \sum_i |c_i|^2\,|a_i\rangle\langle a_i|. \end{aligned} \tag{2.35}$$

The resulting measurement selects between the alternative states $|a_i\rangle$. One may say that it selects between the projectors $P_i|a_i\rangle\langle a_i|$ onto the corresponding one-dimensional subspaces.

Let us now change the realization to obtain a measurement selecting between alternative projectors of more general form. Let the vectors $|a_{i\alpha}\rangle$ with different α but the same i span the subspace $\mathcal{H}_i$ in the state space of the system $\mathcal{S}$. The projector onto this subspace is then

$$P_i = \sum_\alpha |a_{i\alpha}\rangle\langle a_{i\alpha}|.$$

Assume that the evolution of the system $\mathcal{S}+\mathcal{E}$ distinguishes between these subspaces but not inside each of them:

$$|a_{i\alpha}\rangle\,|\Phi_0\rangle \xrightarrow{U} |a_{i\alpha}\rangle\,|\Phi_i\rangle. \tag{2.36}$$

The evolution starting from an arbitrary state

$$|\psi\rangle = \sum_{i\alpha} c_{i\alpha}|a_{i\alpha}\rangle$$

of the system then has the form

$$|\psi\rangle\,|\Phi_0\rangle = \sum_{i\alpha} c_{i\alpha}|a_{i\alpha}\rangle\,|\Phi_0\rangle \xrightarrow{U} \sum_{i\alpha} c_{i\alpha}|a_{i\alpha}\rangle\,|\Phi_i\rangle = |\Psi\rangle. \tag{2.37}$$

The reduced density matrix (calculated with the orthogonality (2.34) taken into account) now has the form

$$\begin{aligned}\rho &= \mathrm{tr}_\Phi |\Psi\rangle\langle\Psi| \\ &= \sum_i |\psi_i\rangle\langle\psi_i|, \end{aligned} \tag{2.38}$$

where

$$\begin{aligned}|\psi_i\rangle &= \sum_\alpha c_{i\alpha}|a_{i\alpha}\rangle \\ &= P_i|\psi\rangle. \end{aligned}$$

This precisely corresponds to the phenomenological scheme (2.19) of the measurement determined by a set of projectors.

We have considered so far only one initial state of the environment leading to the measurement $\{P_i\}$ of the system. Let many initial states exist which lead to a measurement of the same kind, therefore the following correlation

$$|a_{i\alpha}\rangle\,|\Phi_{0m}\rangle \xrightarrow{U} |a_{i\alpha}\rangle\,|\Phi_{im}\rangle.$$

is established for each m. Then for an arbitrary initial state of the system the transformation is of the form

$$|\psi\rangle\,|\Phi_{0m}\rangle = \sum_{i\alpha} c_{i\alpha}|a_{i\alpha}\rangle\,|\Phi_{0m}\rangle \xrightarrow{U} \sum_{i\alpha} c_{i\alpha}|a_{i\alpha}\rangle\,|\Phi_{im}\rangle = |\Psi_m\rangle.$$

The same transformation in terms of density matrices is

$$|\psi\rangle\,|\Phi_{0m}\rangle\langle\Phi_{0m}\,|\langle\psi| \xrightarrow{U} |\Psi_m\rangle\langle\Psi_m|.$$

Taking the trace of the r.h.s. over the environment gives the same reduced density matrix as in Eq. (2.38) for arbitrary m. Summing both sides over m with non-negative coefficients q_m such that $\sum_m q_m = 1$, we have

$$\sum_m q_m|\psi\rangle\,|\Phi_{0m}\rangle\langle\Phi_{0m}\,|\langle\psi| \xrightarrow{U} \sum_m q_m|\Psi_m\rangle\langle\Psi_m|,$$

and the corresponding reduced density matrix will be the same operator ρ of Eq. (2.38). This means that interaction of the system in the state $|\psi\rangle$ with the environment leads to the same density matrix (2.38) not only for a pure initial state $|\Phi_{0m}\rangle$ of the environment, but also for its mixed state described by the density matrix

$$\sum_m q_m|\Phi_{0m}\rangle\langle\Phi_{0m}|,$$

i.e., for an arbitrary mixture of the states $|\Phi_{0m}\rangle$. For any state of the environment from this class the same measurement (presented by the set of projectors $\{P_i\}$) will be performed.

2.3.3. QUANTUM ERASER

We have seen in Sect. 2.2.1 that decoherence is reversible. Eq. (2.12) shows that complete recovery of the initial state of both the measured system and its environment is, in principle, possible. In the case of a simple environment (a meter) this is possible in practice. If we need to recover the state of the measured system but do not care about the state of the environment, then instead of the new period of unitary evolution U' this may be achieved with the help of the next measurement. The device realizing this process is called a *quantum eraser* (SD82; SEW91; KSC92; Hol93). Let us consider how it acts, first for a simple case of two alternatives, then in the general case.

As has been shown in Sect. 2.1, in the case of two alternatives $|a_1\rangle$, $|a_2\rangle$ and the initial state

$$|\psi\rangle = c_1|a_1\rangle + c_2|a_2\rangle,$$

the superposition (the entangled state) forming in the interaction of the system $\mathcal{S}$ and its environment $\mathcal{E}$ is

$$\begin{aligned} |\Psi\rangle &= U|\psi\rangle|\Phi_0\rangle \\ &= c_1|a_1\rangle|\Phi_1\rangle + c_2|a_2\rangle|\Phi_2\rangle. \end{aligned}$$

In simple form the restoration is achieved by the action on this state of the following projector:

$$P = \frac{1}{2}(|\Phi_1\rangle + |\Phi_2\rangle)(\langle\Phi_1| + \langle\Phi_2|).$$

In the resulting state of the system $\mathcal{S} + \mathcal{E}$,

$$P|\Psi\rangle = \frac{1}{2}(|\Phi_1\rangle + |\Phi_2\rangle)|\psi\rangle,$$

the subsystem $\mathcal{S}$ is in the initial state $|\psi\rangle$. The latter is thus restored. The intensity of the signal is decreased because a part of the state $|\Psi\rangle$ is lost. This part is determined by a complementary projector

$$Q = 1 - P = \frac{1}{2}(|\Phi_1\rangle - |\Phi_2\rangle)(\langle\Phi_1| - \langle\Phi_2|)$$

as $Q|\Psi\rangle$. In such a form it does not contain the original state of the system $|\psi\rangle$.. However, the information about the coefficients c_1, c_2 including their relative phase is contained in the state $Q|\Psi\rangle$. As a result the state $|\psi\rangle$ may be obtained with the the help of the additional unitary transformation

$$I = |a_1\rangle\langle a_1| - |a_2\rangle\langle a_2|,$$

changing a sign of $|a_2\rangle$, so that

$$IQ|\Psi\rangle = \frac{1}{2}(|\Phi_1\rangle - |\Phi_2\rangle)|\psi\rangle.$$

Usage of the two complementary projectors P and Q may be organized as a measurement defined by the pair $\{P, Q\}$, on the state $|\Psi\rangle$ (of the compound system $\mathcal{S} + \mathcal{E}$). If the measurement is performed (and therefore its result is known), an additional transformation must be performed: an identical transformation in the case of the first result P and the transformation I in the case of the second result Q. Then, finally, we shall restore the initial state $|\psi\rangle$ of the system $\mathcal{S}$.

The final recipe for the restoration of the initial state ψ of $\mathcal{S}$ is thus: 1) in the state $|\Psi\rangle$ make the measurement $\{P, Q\}$; 2) if the result of the measurement is P, then the state is already restored; 3) if the result is q, apply the transformation I, and the state is restored.

In practice the experiments on a quantum eraser may be realized in the following way (Hol93). The first measurement (of the alternatives $|a_1\rangle$, $|a_2\rangle$) forming the entangled state may be realized by passing a photon along one of the two paths with polarizers in each of them forming vertical or horizontal polarization depending on the path. The entangled state results in correlation between the position of a particle (corresponding to $|a_1\rangle$, $|a_2\rangle$) and its polarization ($|\Phi_1\rangle$, $|\Phi_2\rangle$). Restoration of the initial state is achieved by the next polarizer inducing polarization in the direction of 45^0 to the vertical (and horizontal) direction.

The operation of measurement $\{P, Q\}$ divides the state $|\Psi\rangle$ into two orthogonal pieces $P|\Psi\rangle$ and $Q|\Psi\rangle$, but the complete information about the initial state $|\psi\rangle$ is contained in each of these pieces. This resembles a hologram which also keeps the total information in each part of it. The similarity is enlarged in the following more general case.

Let the environment $\mathcal{E}$ perform the measurement of the system $\mathcal{S}$ described by the complete system of orthogonal projectors $\{P_i\}$. Let the initial state of the system $\mathcal{S}$ be $|\psi\rangle$. Then, according to Sect. 2.3.2, the initial state of the compound system $\mathcal{S} + \mathcal{E}$ is transformed to the following final state:

$$\begin{aligned} |\Psi\rangle &= U|\psi\rangle|\Phi_0\rangle \\ &= \sum_i P_i|\psi\rangle|\Phi_i\rangle. \end{aligned}$$

The simplest quantum eraser may be realized in this case as a projection with the help of the projector

$$P = \frac{1}{N}|\Phi\rangle\langle\Phi|, \quad |\Phi\rangle = \sum_i |\Phi_i\rangle$$

(N being the number of alternatives $\{P_i\}$). It is easily seen that after the projection the state of the compound system is

$$P|\Psi\rangle = \frac{1}{N}|\psi\rangle|\Phi\rangle,$$

therefore the initial state of the system $\mathcal{S}$ is restored.

In this case the initial state is restored from a single 'piece' $P|\Psi\rangle$ of the state $|\Psi\rangle$ (as in a hologram). All the remaining pieces may be used for the restoration too. Let $\epsilon = \{\epsilon_i\}$ be a set of numbers ϵ_i equal to ± 1. Then the projector

$$P_\epsilon = \frac{1}{N}|\Phi_\epsilon\rangle\langle\Phi_\epsilon|, \quad |\Phi_\epsilon\rangle = \sum_i \epsilon_i |\Phi_i\rangle$$

takes the state $|\Psi\rangle$ to the state $P_\epsilon|\Psi\rangle$, from which the sought state $|\psi\rangle$ may be obtained with the help of the additional unitary transformation changing signs of the components:

$$I_\epsilon P_\epsilon|\Psi\rangle = \frac{1}{N}|\psi\rangle|\Phi_\epsilon\rangle, \quad I_\epsilon = \sum_i \epsilon_i P_i.$$

The complete set of operators $\{P_\epsilon\}$ (with arbitrary sets of signs ϵ_i) forms a measurement, and the state $|\psi\rangle$ may be restored in the case of any result of this measurement.

This construction may be generalized even further. Actually almost any projector onto a one-dimensional subspace in the space spanned by the vectors $|\Phi_i\rangle$ may be used as a quantum eraser. Indeed, let

$$P = \frac{|\Phi\rangle\langle\Phi|}{||\Phi||^2}, \quad |\Phi\rangle = \sum_i f_i |\Phi_i\rangle$$

(with $|\Phi_i\rangle$ defined as above). Then the state $|\psi\rangle$ is restored in the following way:

$$IP|\Psi\rangle = |\psi\rangle|\Phi\rangle, \quad I = \sum_i (f_i^*)^{-1} P_i.$$

For an operator I of this form to exist it is enough that none of the coefficients f_i be zero. This requirement is easily interpreted. If at least one of these coefficients is equal to zero, then the state $P|\Psi\rangle$ cannot contain complete information about the initial state $|\psi\rangle$. However, the set of such projectors has measure zero among all projectors in the space spanned by $|\Phi_i\rangle$. The operator I is not, in the general case, unitary, but it may be realized with the help of attenuators and amplifiers. The operation I is unitary when all f_i have unit moduli, and therefore $(f_i^*)^{-1} = f_i$.

2.3.4. REALIZATION OF A FUZZY MEASUREMENT

Let us now show that the description of a measurement by Eq. (2.25) (a fuzzy measurement with a continuous spectrum) may be realized, too.

The following scheme may be used for the realization of this measurement:

	$H_I = gAP$		Meas. of Q	
System	$\longleftrightarrow$	**Meter**	$\longleftrightarrow$	
$A,\ \psi(a)$		$\{\, Q,\ P \,\},\ \Phi(Q)$		

This means that the measured system interacts for a period with a subsidiary system, a meter (which is taken to be one-dimensional with the coordinate Q and linear momentum P), due to the interaction Hamiltonian

$$H_I = gAP,$$

and then the coordinate Q of the meter is precisely measured. We shall see that such a scheme provides a fuzzy measurement of A with fuzziness depending on the initial state of the meter.

Let us assume for simplicity that the period of interaction τ is very short and that the coupling constant g is so large that the interaction Hamiltonian during this period dominates as compared to the free Hamiltonians of the system and the meter. The evolution operator (for the combined system) during the period τ is Then equal to

$$U_\tau = e^{\frac{i}{\hbar}\tau H_I} = e^{-g\tau A \frac{\partial}{\partial Q}}.$$

The operator in the r.h.s. of this formula is nothing other than the displacement operator for the coordinate Q. Therefore as a result of the evolution the coordinate Q will be shifted by a distance proportional to the value of the observable A. More concretely, if at the initial instant the system is in the state

$$|\psi(a)\rangle = \int da\, \psi(a)\, |a\rangle$$

and the meter is in the state

$$|\Phi_0(Q)\rangle = \int dQ\, \Phi_0(Q)\, |Q\rangle,$$

then their evolution during the period τ of interaction will give

$$|\psi(a)\rangle|\Phi_0(Q)\rangle \stackrel{\text{time}\,\tau}{\longrightarrow} \int da\, \psi(a)\, |a\rangle\, |\Phi_0(Q - g\tau a)\rangle.$$

If, now, the measurement of Q is performed giving the result Q', then the vector $|\Phi_0(Q - g\tau a)\rangle$ in the latter state is projected onto $|Q'\rangle$, and we finally have the state

$$\int da\, \psi(a)\, \Phi_0(Q' - g\tau a)\, |a\rangle\ |Q'\rangle.$$

Therefore the combined action of the evolution during τ and the subsequent measurement of Q gives

$$|\psi(a)\rangle|\Phi_0(Q)\rangle \stackrel{\text{time}\,\tau}{\longrightarrow} \stackrel{Q=Q'}{\longrightarrow} R_{a'}|\psi\rangle|Q'\rangle, \tag{2.39}$$

where the notations

$$Q' = g\tau a', \quad R(a - a') = \Phi_0(g\tau(a' - a))$$

are introduced.

We see that in these two steps the state of the system has changed from $|\psi\rangle$ to $R_{a'}|\psi\rangle$, i.e., in accord with the formulas (2.23, 2.25). Therefore the formal description of a fuzzy measurement (2.23) may be realized in the above interaction.

The fuzziness of the measurement (the width of the function $R(a)$) depends in this realization on the uncertainty of the initial state $|\Phi_0(Q)\rangle$ of the meter. This is natural because the result of the measurement a' is determined by the final value Q' of the meter's coordinate. If this coordinate has some uncertainty initially, this uncertainty will reduce the resolution for the determination the value of the observable A.

One more remark must be made about the final step of our procedure, the measurement of Q. This measurement is described formally (according to the projector postulate). In principle, we need some scheme for the realization of this measurement too, but then it will again lead to measurement of some observable, and so on ad infinitum. This is a typical chain leading to the so called 'measurement problem' in quantum mechanics.

It is clear, from the practical point of view, how the difficulty may be resolved. Any real measurement is, in fact, a long chain of interactions and decoherences up to the decoherences in the observer's eye, or further into his brain. However, our task was to understand how the first link in this chain acts. Even if we break the chain somewhere by a formal projection, the physical processes carried out in the first cell of the measuring system (containing the system itself and the meter) may be made physically transparent, and our task efficiently solved. The conceptual problem is not resolved in this way.

We are interested in this book mostly in the technical problems of the correct description of more or less complicated regimes of quantum measurements, and in clarifying the physical meaning of the procedure of measurement (decoherence). The abovementioned 'practical' resolution of the difficulty is quite sufficient for this goal. Some words about conceptual problems will be said at the end of the book.

2.4. Repeated fuzzy measurements as a model of decoherence

We have considered above 'instantaneous' quantum measurements. More precisely, they should be called brief measurements. Actually this means that we neglect the measurement's duration. What happens during this neglected interval is called *decoherence.* We have discussed the physical nature of decoherence in Sect. 2.1. It is interesting that *phenomenologically decoherence may be described as a sequence of instantaneous measurements.*

To illustrate this, consider again the measurement of an observable A with two eigenvalues a_1 and a_2, as in Sect. 2.1. the final result of such a measurement is von Neumann's projection onto one of the eigenvectors. What happens before this may be presented as a sequence of instantaneous fuzzy measurements (Fig. 2.3). Each of these fuzzy measurements has a poor resolution and therefore gives only very poor information about what the value of A is. The influence of such a fuzzy measurement on the system is insignificant. Repetition of this measurement improves the resolution of the eigenvalues, gives more and more definite information about the value of A, and, at the same time, gradually more and more projects the state of the system onto one of the eigenvectors. Thus a long series of instantaneous fuzzy measurements supplies a model of gradual decoherence which leads to von Neumann's projection.

If the state of the system before an instantaneous fuzzy measurement is

$$|\psi\rangle = c_1|a_1\rangle + c_2|a_2\rangle$$

and the fuzzy measurement has a characteristic function $R(a)$, then after the measurement giving the result a' the state of the system is, according to Eq. (2.27),

$$\begin{aligned} |\psi_{a'}\rangle &= R_{a'}\,|\psi\rangle \\ &= R(a_1 - a')c_1|a_1\rangle + R(a_2 - a')c_2|a_2\rangle, \end{aligned}$$

and this result is attained with a probability density equal to

$$p(a') = ||\psi_{a'}||^2 = |\alpha_1|^2|c_1|^2 + |\alpha_2|^2|c_2|^2.$$

The sequence of the measurements is then presented by the following sequence of transformations (see Fig. 2.3):

$$\begin{aligned} c_1|1\rangle + c_2|2\rangle &\rightarrow \alpha_1 c_1|1\rangle + \alpha_2 c_2|2\rangle \\ &\rightarrow \beta_1\alpha_1 c_1|1\rangle + \beta_2\alpha_2 c_2|2\rangle \\ &\ldots\ldots\ldots\ldots\ldots\ldots\ldots \\ &\rightarrow \gamma_1\ldots\beta_1\alpha_1 c_1|1\rangle + \gamma_2\ldots\beta_2\alpha_2 c_2|2\rangle \\ &= \tilde{c}_1|1\rangle \quad \text{or} \quad \tilde{c}_2|2\rangle. \end{aligned}$$

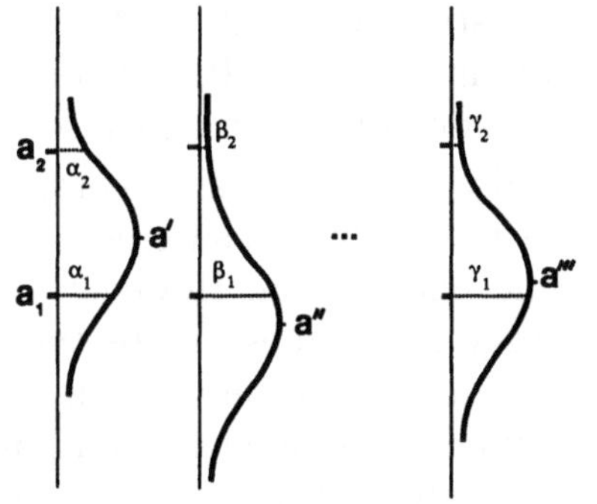

Figure 2.3. Gradual decoherence by a sequence of fuzzy measurements for an observable having two eigenvalues.

This sequence is defined by a series of measurement results $\{a', a'', ...\}$, the probability density of such sequence being equal to the norm of the final vector in Eq. (2.40). Thus we have a stochastic process, and the final state cannot be predicted unambiguously. However, it is clear that the final state may be only $|a_1\rangle$ or $|a_2\rangle$. The reason is that if at some stage of the process the coefficient of $|a_1\rangle$ in the superposition become larger than the coefficient of $|a_2\rangle$, then with high probability it will increase further in the next stage, and so on until the component $|a_2\rangle$ has died out completely. Analogously, the process may converge to $|a_1\rangle$. However, only $|a_1\rangle$ and $|a_2\rangle$ are stable under further repeated measurement.

Therefore the final result of the process may be represented by the von Neumann projection, while the process itself is nothing else than gradual decoherence or continuous fuzzy measurement. In Chapters 7, 8 (see especially Sect. 7.2) we shall consider (with energy as a measured observable) a more realistic description of the process, taking into account not only a sequence of fuzzy measurements but also free evolution of the system between them. We shall then apply another technique, restricted path integrals or complex Hamiltonians, to deal with a continuous rather than repeated measurement. It will be shown in Sect. 7.2 that the probabilities of converging the process to $|a_1\rangle$ or $|a_2\rangle$ are correspondingly $|c_1|^2$ or $|c_1|^2$, as in von Neumann's reduction postulate.

2.5. Conclusion

It cannot be underlined too strongly that the question of quantum measurement is conceptually one of the most complicated (probably, *the* most complicated) in quantum mechanics. Many key papers on quantum measurements, among them those of Bohr and Wigner, were published or reprinted in the collective volume (WZ83). This collection represents very well the great complexity of the question. We shall lightly touch upon the most difficult conceptual problems in Chapter 10, but for now we shall comment only on some technical questions which may cause misunderstanding.

Just as for any actually fundamental theory, radically different approaches to the quantum theory of measurements exist which reflect different aspects of the same phenomena. However, differences in methods, ideology, and language may sometimes create the impression that some of these approaches contradict each other even if they do not. In particular, quantum measurements may be described and investigated at different levels, and it is not immediately clear what are relations between these descriptions.

In the present book we use different levels of description of measurements and try to make their inter-relationships clear. Let us briefly comment on this issue.

If one is only interested in the correct description of the 'practical' aspects of the measurement (alternative results of measurement, their probabilities, and the state of the system after the measurement), it is quite possible to use a phenomenological description including the concept of state reduction (collapse of the wave function). This will never lead to any mistake in practical calculations. From the technical point of view this level of description of the measurement makes use of von Neumann's projectors or their generalizations (see Sect. 2.3). The result of this approach is called the selective description of the measurement.

If one is interested in the statistics of the measurement but not in the behavior of an individual system under the measurement, he may describe the system's state by a density matrix instead of a wave function (a state vector). The concept of collapse is then not necessary. Instead of this, a change of the state in the course of the measurement is described as the conversion of a pure state into a mixed state or (if the initial state is already mixed) as increasing of the degree of mixture. This gives the non-selective description of the measurement. The mathematical techniques applied in this description include the same sets of von Neumann's projectors or their generalizations as in the selective description (see Sect. 2.3).

In the non-selective phenomenological description of a measurement, collapse is not necessary. In the selective description the picture of collapse, or an equivalent non-unitary transformation of the state, is necessary. Yet there is no need, from pragmatic point of view, to ask how and why collapse occurs.

If, however, this question is posed the answer may only be found at a deeper level of description of the measurement. This level must include not only the measured system but also other systems interacting with it and playing the role of a measuring device or measuring medium. An usual quantum mechanical analysis shows (see Sect. 2.1) that the interaction establishes a quantum correlation (or entanglement) between the system and its environment. The system of interest turns out to be a subsystem of a larger system. Therefore its state can not be represented by a wave function, only by a density matrix. This density matrix possesses all properties characteristic of the non-selective phenomenological description of the measurement. In that way the relation is established between the two levels of the description of the measurement.

It is more difficult to understand how the picture of collapse may arise if we consider the interaction of the measured system with its environment. The resulting entangled state contains a superposition of different alternative states of the system in correlation with the corresponding states of the environment. No unitary process caused by an interaction can lead to the disappearance of some components of this superposition. All components exist. From this point of view, collapse is impossible.

However, if the environment contains a macroscopic number of degrees of freedom then the different components of the superposition describe 'macroscopically distinct' states of the environment. Any observer can see only one of these states. The superposition is unobservable (Sect. 2.2). Trying to describe the situation from the point of view of the observer, one is then forced to consider only one component of the superposition. This leads to the picture of collapse which from this point of view is only apparent, not real. Actually, the whole superposition exists, but any observer sees only one of the components of this superposition. Therefore from the point of view of the observer collapse occurs. Even if one knows that all components of the superposition exist, an efficient representation of the point of view of a single observer necessarily leads to the picture of collapse.

Thus different levels of description of the measurement turn out to be quite compatible. None of them can be called wrong if they are understood correctly and in right relation to each other. There are some subtle points in these inter-relations which leave room for the conceptual 'measurement problem', but we prefer to delay their discussion until Chapter 10.

2.6. Comments

The physical nature of quantum measurements has been under consideration since the very beginning. It seems that the fathers of quantum mechanics understood it not so badly (see, for example, the famous book of von Neumann (vN32), the English translation in (vN55)). However, if it existed, at some time this understanding was partly lost. Besides, the real physical questions concerning quantum measurements were closely connected with the so called 'measurement problem', which is, in fact, metaphysical (see Chapter 10).

A mechanism for quantum measurements has been under active consideration several times, and these active periods were divided by decades. Not pretending to give a complete list of literature on the question, we may mention how the question is analysed in the famous book of Bohm (Boh52) and the lectures of Mandelstam (Man72). The modern period of the investigation of the topic began probably with the papers of Zeh and Joos (Zeh70; Zeh73; Joo84; JZ85) and finished with Zurek (Zur81; Zur82). An original 'many Hilbert space' representation of a macroscopic apparatus was developed by Machida and Namiki (MN80a; MN80b) for explanation of collapse. The so called 'consistent histories' technique was applied to relevant questions independently by Griffiths, Omnès, and Gell-Mann & Hartle (Gri84; Omn90; GMH90; GMH93; Omn94) (see, in this connection, (PZ93)). The contemporary state of the whole question is discussed from different points of view in the book of Giulini, Joos, Kiefer, Kupsch, Stamatescu and Zeh (GJK$^+$96).

Another aspect of quantum measurements is their practical realization. It is closely connected with the mechanism of measurement, but we do not seriously consider this aspect here. Let us mention, nevertheless, a good book of Braginsky and Khalili (BK92) devoted just to this question and the book of Peres which is of a more general character (Per93) but containing analyses of many models of measurements. Many examples of quantum measurements realized by means of neutron interferometry as well as experimental demonstration of the basic principles of quantum mechanics may be found in the review of Rauch (Rau98) and the book of Rauch and Werner (RW00).

We have presented in Sect. 2.3, in a very simple manner, the main mathematical models of quantum measurements. A much more complete review of this topic, but in a much more complicated mathematical form, may be found in (Dav76; Hol82; Kra83; BLM91). In particular, the mathematical concepts of 'instrument', 'operation' and 'effect' proposed by Ludwig

(Lud83) are used, in this mathematically-oriented literature, for the description of quantum measurements. The literature of this cathegory also contains very general classes of models of quantum measurements, but also usually in a rather abstract form.

We apply a simpler mathematical formalism which has the advantage of being transparent and, nevertheless, sufficient in most situations. The generality and mathematical subtleties are, of course, lost in this simple presentation. The readers interested in these aspects may appeal to the mathematical books cited above and references therein.

In almost all concrete examples in this book we consider the measurements which were earlier called 'measurements of the first kind' and also their direct generalizations. The measurements of the first kind are described by projectors.[9] This is why the result of the measurement will be the same if such a measurement is repeated just after the first measurement is over. The direct generalization is the measurement which changes the system state by multiplying it by a positive operator instead of a projector. It is more realistic.

Both these types of measurements may be called 'minimally disturbing'. They allow one to obtain certain information about the system with only minimally disturbing its state (only the modulus, and not phase, of the wave function is changed). This class of measurements is actually the most important, although the measurements changing the phase relations may be advantageous in some specific situations.

[9]although even this, strictly speaking, needs to be made more precise, see (BLM91)

Chapter 3

Measurement induced uncertainties

The well known Heisenberg uncertainty relations (UR) concern values of corresponding observables in a given state $|\psi\rangle$. However, UR may be applied also to the situation of measurement. In this case the uncertainties included in the UR have a quite different sense. The UR induced by measurement inter-relate the measurement error (resolution) and the uncertainty of the conjugate observable in the state resulting in the measurement.

We shall formulate and prove these measurement UR in two simple forms and then consider their application to the phenomenon of the loss of the interference pattern in a measurement. In this connection we shall consider so called 'Which Way' experiments, starting from the classical two-slit experiment and finishing with the modern experiment with the 'which way' information recorded in the internal state of the particle (atom).

3.1. Measurement induced uncertainty relation

The Heisenberg UR for the coordinate and linear momentum have the form (see Sect. A2)

$$\Delta p \Delta q \geq \frac{\hbar}{2}, \tag{3.1}$$

where the square of the uncertainty (variance) of an operator A is defined as follows:

$$\begin{aligned} \Delta A^2 &= \langle\psi|(A - \langle\psi|A|\psi\rangle)^2|\psi\rangle \\ &= \langle\psi|A^2|\psi\rangle - \langle\psi|A|\psi\rangle^2. \end{aligned} \tag{3.2}$$

More generally, if the two Hermitian operators X, Y are canonically conjugate to each other (i.e., their commutation relation is the same as for the coordinate and momentum), then

$$\Delta X \Delta Y \geq \frac{\hbar}{2}. \tag{3.3}$$

Even more generally, for two arbitrary Hermitian operators A, B the following UR is always valid (see Sect. A2):

$$\Delta A \Delta B \geq \frac{1}{2}|\langle\psi|[A, B]|\psi\rangle|. \tag{3.4}$$

UR may be applied also to the situation of measurement. In this case the uncertainties included in the UR have a quite different sense. Let us consider this question.

3.1.1. FORMULATION OF THE MEASUREMENT UR

The following statement is often used in the quantum theory of measurements: if the coordinate q is measured with a finite precision, then the linear momentum p obtains, as a result of the measurement, an additional uncertainty determined by the UR. This statement deals with the two different states (before and after the measurement) and may be illustrated by the following scheme:

$$\{\Delta q, \Delta p\} \stackrel{Dq}{\to} \{\delta q, \delta p\}, \quad \delta p \geq \frac{1}{2}\frac{\hbar}{Dq}. \tag{3.5}$$

The uncertainties Δq, Δp characterize the state before the measurement, δq, δp refer to the state after the measurement, Dq is the measurement error (resolution). The statement claims that Dq and δp are connected by the UR. In other words, if q is measured with a resolution Dq, then the resulting uncertainty of p is not less than $\hbar/2Dq$.

For an arbitrary pair of canonically conjugate observables X, Y an analogous statement should be valid:

$$\{\Delta X, \Delta Y\} \stackrel{DX}{\to} \{\delta X, \delta Y\}, \quad \delta Y \geq \frac{1}{2}\frac{\hbar}{DX} \tag{3.6}$$

3.1.2. PROOF OF THE MEASUREMENT UR

To prove the statement (for definiteness, for q and p) we shall describe the measurement of the coordinate by the scheme analogous to (2.23):

$$\psi(q) \stackrel{x}{\to} R(q-x)\psi(q) = \psi_x(q). \tag{3.7}$$

The initial state is described by the wave function ψ in the q-representation. The final state depends on the result x of the measurement of the coordinate and is determined by the function $R(q)$, a characteristic of the measurement. We shall assume that $R(q)$ is a 'bell shape' positive function, though all the results will be valid also in generic case.[1] The width of this function determines the resolution of the measurement. The probability density of the measurement result x is equal to the norm of the final state as it is defined in Eq. (3.7):

$$p(x) = ||\psi_x||^2. \tag{3.8}$$

[1]If $R(q)$ is not positive $R^2(q)$ must be replaced by $|R(q)|^2$ in all formulas, which does not change the arguments.

This definition is illustrated by Fig. 3.1. The two vertical columns of

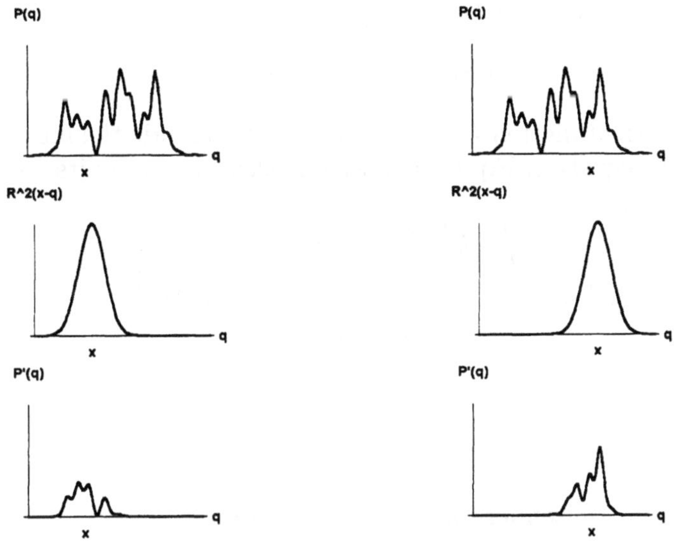

Figure 3.1. Smooth projecting of the wave function in a fuzzy measurement of the coordinate. The initial wave function (upper row of diagrams) converts into the final function (lower row) obtained by multiplication with the characteristic function of the measurement. The resulting wave function depends on the output of the measurement x (left or right hand vertical column of diagrams).

diagrams in this figure present two different measurement results x. Depending on this result, the initial wave function (represented in the upper diagrams by the function $P(q) = |\psi(q)|^2$) is converted after the measurement into one of the wave functions of the lower diagrams.

Let the uncertainties in the initial state ψ be Δq and Δp. The uncertainties in the final state ψ_x depend on x, but at the moment we shall denote them simply by δq and δp, not including x (in Sect. 3.1.3 we shall be forced to introduce more accurate notations).

To estimate the uncertainty δq let us make use of the relations (see Sect. A2)

$$\delta q^2 = \overline{(q-\overline{q})^2} = \min_a \overline{(q-a)^2} \leq \overline{(q-x)^2}$$

(with averaging over the state ψ_x) or, in a more explicit form,

$$\delta q^2 \leq \frac{\int (q-x)^2 R^2(q-x)|\psi(q)|^2}{\int R^2(q-x)|\psi(q)|^2} = \frac{\int q^2 R^2(q)|\psi(q+x)|^2}{\int R^2(q)|\psi(q+x)|^2}. \tag{3.9}$$

We shall now take into account that the measurement's resolution is equal to Dq. This means that the width of the function $R(q)$ is Dq. The width may be defined in different ways. To make the proof simpler we shall assume for the moment that the function $R(q)$ is symmetric and has a restricted support of width $2Dq$:

$$\text{supp}\, R \subset [-Dq, Dq]. \tag{3.10}$$

This means that the function $R(q)$ is equal to zero outside the interval [-Dq, Dq]. Then the integrals in Eq. (3.9) may be taken in the limits $[-Dq, Dq]$:

$$\delta q^2 \leq \frac{\int\limits_{|q|\leq Dq} q^2 R^2(q)|\psi(q+x)|^2}{\int\limits_{|q|\leq Dq} R^2(q)|\psi(q+x)|^2} \leq Dq^2$$

This gives

$$\delta q \leq Dq.$$

From the usual Heisenberg uncertainty relation

$$\delta q\, \delta p \geq \frac{\hbar}{2},$$

and taking into account the preceding inequality, we have finally the measurement UR:

$$\delta p \geq \frac{\hbar}{2\delta q} \geq \frac{\hbar}{2Dq}. \tag{3.11}$$

We have proved the required uncertainty assuming that the result of the measurement x is precisely known. Strictly speaking, this is impossible because the spectrum of x is continuous. However, the result is also approximately valid if it is known that the measurement result x lies in an interval of width $\Delta x \ll Dq$. A precise formulation of this statement is the following.

If the measurement result is known with some uncertainty, the state of the system after the measurement must be described in a partly nonselective way as in Eq. (2.32). Let, in our case, the final state be represented by the density matrix

$$\rho = \int dx\, p(x)\, \frac{|\psi_x\rangle\langle\psi_x|}{||\psi_x||^2}, \quad \int dx\, p(x) = 1.$$

If the result is known with uncertainty Δx then the width of the function $p(x)$ should be Δx. Assume that the support of the function $p(x)$ is restricted and has the width $2\Delta x$:

$$\text{supp}\, p \subset [a - \Delta x, a + \Delta x]. \tag{3.12}$$

Then the final uncertainty of the linear momentum satisfies the inequality

$$\delta p_\rho \geq \frac{\hbar}{2(Dq + \Delta x)}.$$

The uncertainty is defined here as

$$\delta p_\rho^2 = \langle (p - \langle p \rangle_\rho)^2 \rangle_\rho$$

with the notation

$$\langle A \rangle_\rho = \mathrm{tr}(\rho A).$$

The proof follows from the uncertainty relation for the mixed state

$$\delta p_\rho \delta q_\rho \geq \frac{\hbar^2}{4},$$

the inequality

$$\delta q_\rho^2 = \langle (q - \langle q \rangle_\rho)^2 \rangle_\rho \leq \langle (q - a)^2 \rangle_\rho$$

(valid for arbitrary a) and the estimate

$$\int dx \, \frac{p(x)}{||\psi_x||^2} \int dq \, (q-a)^2 \, R^2(q-x) \, |\psi(q)|^2 \leq (Dq + \Delta x)^2$$

following from Eqs. (3.10, 3.12).

There is no need to say that the above proof is straightforwardly generalized to the case of an arbitrary pair X, Y of canonically conjugate variables. In particular, if the momentum is measured with resolution Dp we have the following relation for the uncertainty of the coordinate:

$$\{\Delta q, \Delta p\} \stackrel{Dp}{\to} \{\delta q, \delta p\}, \quad \delta q \geq \frac{1}{2} \frac{\hbar}{Dp}. \tag{3.13}$$

Starting from the more general Heisenberg UR (3.4), the measurement uncertainty for an arbitrary pair A, B of Hermitian operators may be proved. If the commutator of the operators is

$$[A, B] = i\,C,$$

then

$$\{\Delta A, \Delta B\} \stackrel{DA}{\to} \{\delta A, \delta B\}, \quad \delta B \geq \frac{1}{2} \frac{|\langle C \rangle|}{DA}, \tag{3.14}$$

where the mean value of C in the final state (after the measurement) must be substituted. This mean value depends on the measurement result, so the general measurement UR is not so convenient as for a c-number commutator.

In the preceding we have assumed that the characteristic function $R(q)$ of the measurement has a restricted support, i.e., is precisely null outsize the interval $[-Dq, Dq]$. This assumption radically simplifies the proof, but it is not realistic. In reality the function $R(q)$ may gradually die out but be nonzero far beyond the region where it has significant values. This is the case for example for the Gaussian characteristic function. The question naturally arises of whether the measurement uncertainty relation is applicable in the corresponding measurements or not.

An evident answer is the following. A realistic function $R(q)$ may be replaced by another one coinciding with $R(q)$ in the region where the latter has significant values but is precisely zero outside this region. Then the formulated measurement UR (3.11) may be considered as a good approximation for the real situation.

Yet it is possible to give a more rigorous and physically more precise solution of the problem as follows.

3.1.3. MORE PRECISE FORM OF THE MEASUREMENT INDUCED UR

Let the measurement's resolution be determined by the width of the characteristic function $R(q)$. Denote it Dq and define, instead of Eq. (3.10), by

$$(Dq)^2 = \int q^2\, R^2(q)\, dq. \tag{3.15}$$

This definition is constructed in analogy with the concept of the coordinate's uncertainty (notice that the function $R(q)$ is normalized according to (2.26)). With this definition, the support of the function $R(q)$ may be much wider than of the width Dq. It may even be infinitely wide. Now consider the measurement (3.7) with the characteristic function $R(q)$ having these properties. Our task is to analyse the momentum uncertainty δp_x in the state $\psi_x(q)$ resulting in the measurement.

If the support of the function $R(q)$ is large, then the coordinate's uncertainty δq_x in the state $\psi_x(q)$ may also be large (compared with Dq) for some x. Therefore despite the Heisenberg UR

$$\delta q_x\, \delta p_x \geq \frac{\hbar}{2}, \tag{3.16}$$

the uncertainty δp_x may be small (in comparison with $\hbar/2Dq$). However, this happens only for those values x which have very small probability densities and therefore practically never arise as measurement results.[2]

[2] As an example, let $\psi(q)$ consist of two equal peaks located far from each other symmetrically about $q = 0$ and the characteristic function $R(q)$ be Gaussian. Then the most probable values of x are those almost coinciding with the locations of the peaks. In

To express this in a more precise way we shall average (in an appropriate way) different values of δp_x with the probabilities of the corresponding values x taken into account, then the values δp_x larger than $\hbar/2Dq$ will give a small contribution, so the UR will be valid after averaging.

Recall that the probability for x to be a result of the measurement is equal to the square norm $||\psi_x||^2$ of the final state of the system (see Eq. (3.8)). The measurement induced uncertainty relation may be written in the following form:

$$\int \frac{||\psi_x||^2}{\delta p_x^2}\, dx \leq \frac{4}{\hbar^2} Dq^2. \tag{3.17}$$

To prove this, write down the following chain of inequalities beginning with the Heisenberg UR (3.16) (see Sect. A2 for proof of the last inequality)

$$\begin{aligned} \frac{\hbar^2}{4}\frac{1}{\delta p_x^2} &\leq \delta q_x^2 = \min_a \frac{\langle\psi_x|(q-a)^2|\psi_x\rangle}{||\psi_x||^2} \\ &\leq \frac{\langle\psi_x|(q-x)^2|\psi_x\rangle}{||\psi_x||^2}. \end{aligned}$$

Now represent the mean value in the r.h.s. in an explicit form (an integral over q). This gives

$$\frac{\hbar^2}{4}\frac{||\psi_x||^2}{\delta p_x^2} \leq \int (q-x)^2\, R^2(q-x)\, |\psi(q)|^2\, dq.$$

Integrating this inequality over x, assuming that the function $R(q)$ is symmetric and taking into account the definition (3.15), we have finally Eq. (3.17).

3.2. Washing interference out

The uncertainty relation (3.5) for measurement of the coordinate is always valid, but it is especially efficient if the initial uncertainty of the linear momentum is smaller than the uncertainty arising in the measurement, $\Delta p \ll \Delta p_{\text{meas}}$ where

$$\Delta p_{\text{meas}} = \frac{\hbar}{2Dq}.$$

In this case the final uncertainty δp is determined mostly by Δp_{meas}.

The situation is more complicated in the opposite case, $\Delta p \gg \Delta p_{\text{meas}}$. In this case the uncertainty Δp_{meas} induced by the measurement will be

all these cases the coordinate uncertainties δq_x will be small (of the order of the width of the peaks) and δp_x large, satisfying the measurement UR. However, x may be close to zero, although with exponentially small probability. Then δq_x will be equal to the distance between the peaks, i.e., very large.

added to the initial uncertainty Δp. It may seem that this addition may have no influence if $\Delta p \gg \Delta p_{\rm meas}$. This naïve judgement is, however, invalid because the additional uncertainty may radically change the structure of the momentum distribution. It is just because of this effect that an interference pattern may be lost (washed out) as a result of a measurement. To analyse this effect we need to know how the momentum distribution is changed by an additional uncertainty acquired in the measurement of the coordinate.

3.2.1. ADDING THE MEASUREMENT UNCERTAINTY

Consider the momentum representation of the initial state ψ:

$$\widetilde{\psi}(p) = \frac{1}{\sqrt{2\pi\hbar}} \int e^{-\frac{i}{\hbar}pq} \psi(q)\, dq.$$

As a result of the measurement of the coordinate,

$$\psi(q) \xrightarrow{x} R_x(q)\psi(q) = \psi_x(q), \quad R_x(q) = R(q-x),$$

we have a new state ψ_x. It is easily proved that its momentum representation has the form

$$\widetilde{\psi}_x(p) = \int \widetilde{R}_x(p')\, \widetilde{\psi}(p-p')\, dp' = \int \widetilde{R}_x(p-p')\, \widetilde{\psi}(p')\, dp', \qquad (3.18)$$

where $\widetilde{R}(p)$ is a 'momentum representation' of the function $R(q)$ (but with another numerical factor in front of the integral)

$$\widetilde{R}(p) = \frac{1}{2\pi\hbar} \int e^{-\frac{i}{\hbar}pq} R(q)\, dq,$$

and $\widetilde{R}_x(p)$ differs only by a phase factor:

$$\widetilde{R}_x(p) = e^{-\frac{i}{\hbar}px} \widetilde{R}(p).$$

The function $R(q)$ is a characteristic function of the measurement of the coordinate. From the above formulas it is clear that the function $\widetilde{R}(p)$ presents the structure of the uncertainty acquired by the system in the measurement, while Eq. (3.18) gives the law of 'addition' this uncertainty to that existing before the measurement (STCW94). The 'addition' occurs, according to this law, by the operation of *convolution*:

$$(f * g)(x) = \int f(y)\, g(x-y)\, dy = \int f(x-y)\, g(y)\, dy \qquad (3.19)$$

(the limits of integration are assumed infinite).

The physical nature of the additional uncertainty in p is the momentum transfer from the environment to the measured system performed in the process of interaction between them. In this connection the momentum representation $\widetilde{R}(p)$ of the measurement characteristic function $R(q)$ has been called in (STCW94) 'momentum kick amplitude distribution'. However, it is important to be careful with the term 'kick' to avoid its classical treatment. The correct treatment is given by the convolution mechanism.

The convolution mechanism may wash out a periodic structure in the initial distribution. Indeed, Eq. (3.18) may be represented in the form

$$e^{\frac{i}{\hbar}px}\widetilde{\psi}_x(p) = \int \widetilde{R}(p')\, e^{\frac{i}{\hbar}(p-p')x}\widetilde{\psi}(p-p')\, dp' = \int \widetilde{R}(p-p')\, e^{\frac{i}{\hbar}p'x}\widetilde{\psi}(p')\, dp'.$$

The function $\widetilde{R}(p)$ in the integrand is a bell-like function of width $\Delta p_{\text{meas}} = \hbar/2Dq$. Therefore the function in the l.h.s. cannot contain Fourier components with periods less than Δp_{meas}. The same is valid also for its square modulus $|\widetilde{\psi}_x(p)|^2$.

Now let the momentum representation of the initial state $\widetilde{\psi}(p)$ be 'amplitude-modulated', i.e., the function $|\widetilde{\psi}(p)|^2$ is close (in some interval of the argument) to a periodic function with period Λ. If Δp_{meas} is greater than Λ then after the convolution (3.18) the periodic structure (the amplitude modulation) will disappear. A periodic structure usually means an interference pattern, so the interference pattern is also washed out as a result of the measurement.

3.2.2. EXAMPLE: WASHING OUT THE PERIODIC STRUCTURE

Consider a simple example in which a periodic structure existing in the momentum distribution of the initial state is washed out by the measurement of the coordinate.[3] We shall omit numerical factors in the functions to be evaluated.

Consider a coordinate measurement having a Gaussian characteristic function of width of the order of Dq:

$$R(q) \sim e^{-q^2/4Dq^2}.$$

Then the 'momentum representation' of this function (characterizing the structure of the momentum uncertainty acquired in the measurement) is also Gaussian with width of the order of $\hbar/Dq$:

$$\widetilde{R}(p) \sim e^{-p^2/(\hbar/Dq)^2}.$$

[3]This example is, in fact, a simple version of the two-slit Which Way experiment to be considered in Sect. 3.3.

Let the initial state be a superposition of two very narrow peaks separated by the distance d:

$$\psi(q) \sim \delta(q - d/2) + \delta(q + d/2). \tag{3.20}$$

Then the momentum representation of this state is

$$\widetilde{\psi}(p) \sim \cos\frac{pd}{2\hbar}, \quad |\widetilde{\psi}(p)|^2 \sim 1 + \cos\frac{pd}{\hbar}.$$

The momentum distribution in this state has a periodic structure with period $2\pi\hbar/d$. It may be interpreted as an interference pattern (in the far zone) caused by the waves from the two positions of the system (the particle) in the points $q = \pm d/2$ (this may be a mathematical representation of the two-slit experiment, see Sect. 3.3).

The momentum representation of the state after the measurement will be

$$\widetilde{\psi}_x(p) \sim \exp\left[-\frac{(x+d/2)}{(2Dq)^2} - \frac{i}{2\hbar}pd\right] + \exp\left[-\frac{(x-d/2)}{(2Dq)^2} + \frac{i}{2\hbar}pd\right]. \tag{3.21}$$

The periodic structure is washed out provided that

$$Dq \ll d, \tag{3.22}$$

i.e., if the coordinate is measured with an error much less than the distance between two locations of the particle. To prove that the periodic structure disappears in this case, we also need the norm of the vector ψ_x to evaluate the probability density of the measurement to produce the result x. It may be shown that the probability density is[4]

$$p(x) = \frac{1}{2Dq\sqrt{2\pi}}\left[e^{-\frac{(x-d/2)^2}{2Dq^2}} + e^{-\frac{(x+d/2)^2}{2Dq^2}}\right]$$

Under the condition (3.22) the probability density is not negligible only for $x \approx \pm d/2$. This is natural because the measurement is precise enough and cannot give a result far from the locations of the particle. Substituting $x \approx \pm d/2$ in Eq. (3.21) we see that one of the two terms in the r.h.s. is negligible, so the periodic structure disappears.

Interpretation of this property is evident. If the measurement is precise enough to resolve between the two peaks in the coordinate representation,

[4]An evident difficulty arises because the function (3.20) is unnormalizable. The way out of the difficulty is that in the transition from ψ to ψ_x the infinite norm of the function acquires an additional finite factor, its square being the probability density. More strictly speaking, one may replace the sum of two delta functions (3.20) by the sum of two peaks, narrow but of finite width, as in Sect. 3.3.

then only one of these peaks survives after the measurement. Therefore the interference between the two peaks is lost as a result of the measurement.

3.3. Two-slit Which Way experiment

An example of washing the interference pattern out, analysed in Sect. 3.2, is in fact a mathematical treatment of the Which Way experiment on the basis of the famous two-slit experiment, the best known illustration of features of quantum mechanics. The question of interference patterns lost as a result of measurements (Which Way experiments) was recently intensively discussed (SEW91; ESW95; STCW94; WHC$^+$97; WH95; DNR98). We shall consider this question here and in Sect. 3.4 beginning with the classic two-slit experiment and finishing by its modern quite sophisticated analogue (DNR98).

A Which Way experiment on the basis of the classic two-slit setup may be arranged in various ways, for example with the help of the Einstein's recoiling slit (Boh83) or Feynman's light microscope (FLS65, Volume III, Chapter I). However, the physical content of the experiment is always the same. It is schematically presented in Fig. 3.2 (see also Sect. 5.1.1). The

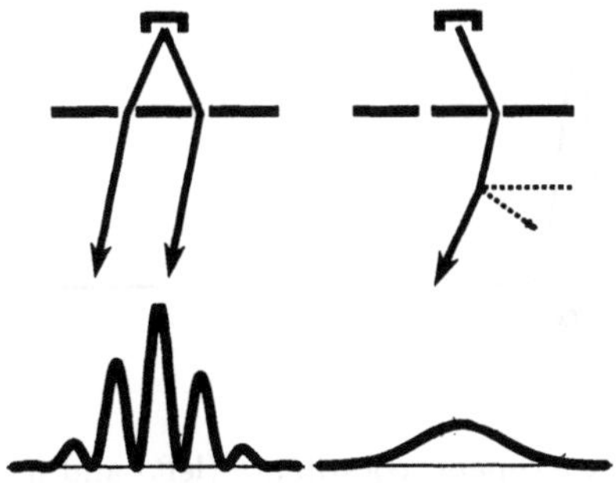

Figure 3.2. The two-slit experiment: (a) an interference pattern arises if no measurement is performed of the path of the particle; (b) it disappears if such a measurement is performed.

interference pattern arises if there is no information about the way chosen by the particle (the left part of Fig. 3.2). If the 'which way' information is obtained from some measurement the interference pattern disappears (the right part of Fig. 3.2). The measurement may be arranged in various ways, but always may it be interpreted as a measurement of the position of the particle (the position in horizontal direction at the moment of passing the opaque screen with the slits) with a precision sufficient for resolving the two slits.

If we denote the horizontal coordinate by q and the corresponding (horizontal) linear momentum by p, then the state of the particle at the moment of passing the screen is described by the following wave function in the coordinate and momentum representations (we omit the normalization factors):

$$\begin{aligned}\psi(q) &\sim e^{-(q-d/2)^2/4\Lambda^2} + e^{-(q+d/2)^2/4\Lambda^2},\\ \widetilde{\psi}(p) &\sim e^{-(\Lambda^2/\hbar^2)p^2}\cos\frac{pd}{2\hbar}, \quad |\widetilde{\psi}(p)|^2 \sim \frac{1}{2}e^{-2(\Lambda^2/\hbar^2)p^2}\left(1+\cos\frac{pd}{\hbar}\right).\end{aligned}$$

This state is shown in the upper part of Fig. 3.3. Two characteristic peaks

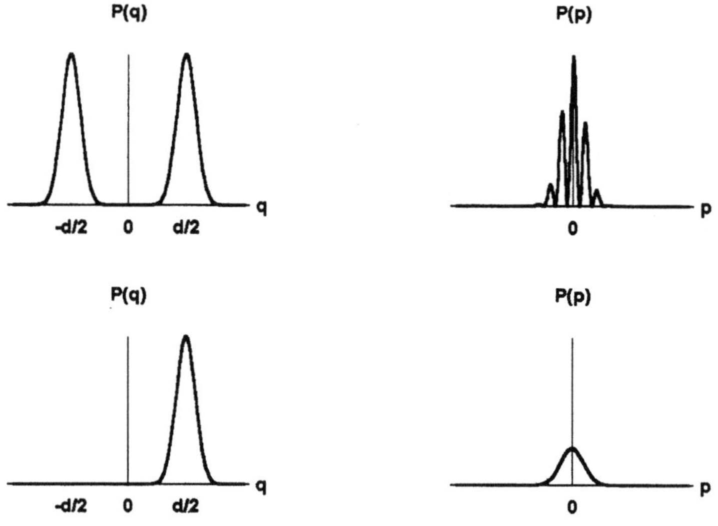

Figure 3.3. The two-slit experiment: coordinate (left) and momentum (right) representations of the state localized near a single point $d/2$ (lower row) or distributed near two points $-d/2$ and $d/2$ (upper row)

in the q-representation correspond to the locations of the particles in the two slits. The p-representation has a characteristic periodic structure. This means that the deflection of the particle in the horizontal direction contains interference fringes which may be observed by a detector in the far zone (see Fig. 3.2). This situation pertains if no measurement supplying the 'which way' information is performed.

If the measurement is performed then it is known (or at least may be found, in principle) which slit is chosen by the particle. Then the state of the particle is described by a wave function having only one peak in the q-representation and no periodic structure in the p-representation (the lower part of Fig. 3.3):

$$\psi'(q) \sim e^{-\frac{(q-d/2)^2}{4\Lambda^2}}, \quad \tilde{\psi}'(p) \sim \frac{1}{2} e^{-\frac{\Lambda^2}{\hbar^2}p^2 + i\frac{pd}{2\hbar}}, \quad |\tilde{\psi}'(p)|^2 \sim \frac{1}{4} e^{-2\frac{\Lambda^2}{\hbar^2}p^2}.$$

These formulas may be elaborated by the method expounded in detail in Sect. 3.2. An example given in Sect. 3.2.2 is exactly the same as that discussed here but with less precise approximation when the widths of the slits are negligible.

Recalling what has been said in Sect. 3.2 about measurement induced uncertainties, we may formulate the following interpretation of the above formulas. In the usual (without the 'which way' measurement) scheme of the two-slit experiment the p-representation of the wave function has periodic structure leading to the interference pattern in the far zone. If the which way information is found in some way or other, this is equivalent to the measurement of the coordinate (resolving between the slits). Such a measurement induces an additional uncertainty in the momentum p, which by the convolution mechanism (Sect. 3.2.1) washes out the periodic structure in the p-representation. The interference pattern disappears.

The physical nature of the additional uncertainty in p is, in fact, the momentum transfer in the process of the interaction of a particle with the device which measures the coordinate. However, it is important to avoid the classical consideration of this transfer. Instead of this, the convolution mechanism must be applied as it is argued in (Sect. 3.2.1).

3.4. Recording Which Way information in the internal state

In the paper (DNR98) a Which Way experiment with an beam of atoms interfering after being split by two 'optical lattices' is described and interpreted. If the information about the direction taken by a given atom (which way information) is recorded in its internal state the interference pattern disappears.

The authors of (DNR98) argue that the disappearance of the interference pattern is a resuslt of the entanglement between the atom's position and internal state but can not be explained by the uncertainty principle as it is usually done in the cases of Einstein's recoiling slit (Boh83) and Feynman's light microscope (FLS65, Volume III, Chapter I). In (Kni98) this statement is (at least partly) supported. This discussion is, in fact, a step in the long controversial polemic (SEW91; ESW95; STCW94; WHC+97; WH95) about whether complementarity (particle–wave duality) is equivalent to the uncertainty principle or is more general.

We will demonstrate in Sect. 3.4.3 that the explanation of the given experiment on the basis of the uncertainty principle is possible, but with the roles of the coordinate and momentum interchanged: momentum is measured and the coordinate acquires additional uncertainty.

The question is more complicated in the modification of this experiment realized by the authors but not mentioned in (DNR98). We shall consider this modification too. In this case the phenomenon is more complicated, but nevertheless it can be described by the measurement uncertainty relation. To make the essential issues more clear we shall discuss the situation in much detail and from different points of view. At the end of Sect. 3.4.4 a simple formal solution of the problem will be given: the recording of the which way information will be presented as a measurement of a certain observable, and the resulting uncertainty will be shown to wash out the interference pattern.

Let us emphasise that the interference is lost in this experiment as a result of decoherence caused by entanglement just as it could be lost in an irreversible measurement. Irreversibility is not a principal point, see Sect. 2.2.1.

3.4.1. EXPERIMENTAL SETUP

The experimental setup discussed in (DNR98) includes a beam of atoms in the state $|2\rangle$ crossing the two standing waves of light, 1 and 2, and detected afterwards (see Fig. 3.4 a). Each standing wave is a sort of 'optical

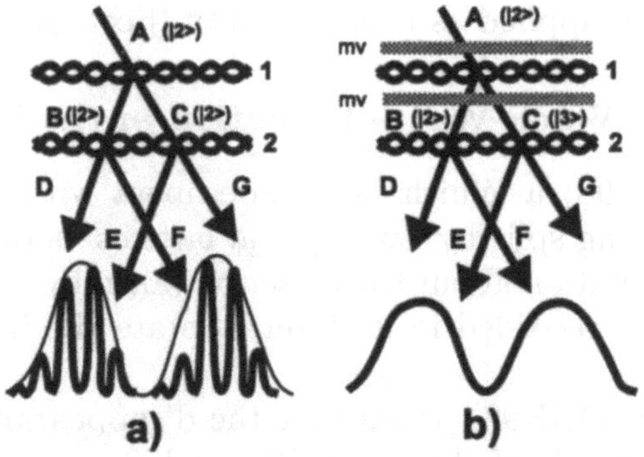

Figure 3.4. Interferometer with an atom beam, two optical interference lattices (standing waves of light) and a recorder of which way information: (a) an interference pattern arises if the recorder is turned off; (b) no interference pattern exists if the recorder is turned on (the radiation is turned on in the two microcavities sandwiching the first splitter).

interference lattice' and acts as a beam splitter: an atom may be transferred forward or deflected by the lattice (the Bragg reflection). The two beam splitters 1 and 2 form, finally, four coherent beams giving two interference patterns at the two detectors, one for the atoms finally deflected to the right and the other for the atoms deflected to the left. The interference patterns disappear (Fig. 3.4 b) if the 'which way recorder' constructed in the following way is turned on.

The first standing wave is sandwiched by two microcavities (Fig. 3.5 b) which enable the experimentalist to record or not record the which way information in the internal state of the atoms.[5] The radiation in the mi-

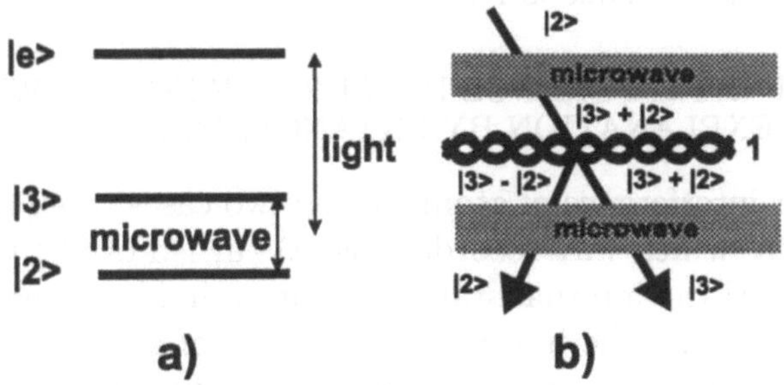

Figure 3.5. Recording which way information: a) Light in the standing waves 1 and 2 is tuned halfway between the $|2\rangle \rightarrow |e\rangle$ and $|3\rangle \rightarrow |e\rangle$ transitions. b) This results in the change of the sign of the state $|2\rangle$ during the reflection by the standing wave 1. The $\pi/2$-pulses of the microwaves sandwiching the standing wave 1 (in resonance with the $|2\rangle \rightarrow |3\rangle$ transition) transfer the atom to the state $|2\rangle$ or $|3\rangle$ depending on whether it is reflected or not by the splitter.

crocavities is in resonance with the transition between the internal states $|2\rangle$ or $|3\rangle$. The duration of the radiation in each of these microcavities corresponds to the $\pi/2$-pulse. Light in the standing wave forming the optical lattice is tuned in such a way that its detuning with respect to the transition $|3\rangle - |e\rangle$ is positive and the detuning with respect to the transition $|2\rangle - |e\rangle$ is negative (Fig. 3.5 a) for some excited level $|e\rangle$. Under this condition the Bragg reflection of the atom from the optical lattice occurs with a loss of phase π if the atom is in the state $|2\rangle$ and without a loss of the phase for the state $|3\rangle$.

Therefore splitting the beam of atoms at the first beam splitter is accompanied by the change of each atom's internal state (initially $|2\rangle$) as is shown in Fig. 3.5 b if the radiation in the microcavities is turned on. If it is turned off the internal state of atoms does not change and remains equal to the initial state $|2\rangle$.

Taking into account both beam splitters, we have, for the radiation in the microcavities turned on, a change of the internal states as shown in Fig. 3.4 b. The internal state of the atom depends in this case on which of the channels B or C is taken by it after it passes beam splitter 1. Namely, atoms deflected by beam splitter 1 (beam B) are in the state $|2\rangle$ while those

[5]In another, and even more interesting variant of the experiment the second standing wave is sandwiched instead of the first. We shall discuss this variant later, but for now will consider in detail the first one.

transferred without deflection (beam C) are in the state $|3\rangle$ (see (DNR98) for details). Therefore if the radiation in the microcavities is turned on the beams B and C become distinguishable and do not interfere. This leads to the loss of the interference pattern in the case of the radiation being turned on. If the radiation is turned off (Fig. 3.4 a) the which way information is absent and the interference is restored.

3.4.2. LOSS OF INTERFERENCE PATTERN OWING TO 'WHICH WAY' RECORDING (EXPLANATION BY ENTANGLEMENT)

Let us consider interference of atoms in the two cases when radiation in the microcavities (a 'which way recorder') is: a) turned off; b) turned on.

In the case (a) the internal state of an atom in all beams is equal to the initial state. Denoting the initial internal state of an atom $|2\rangle$ we have for the state of the atom after passing through the first beam splitter

$$|\psi_1\rangle = -|\psi_{\rm B}\rangle \otimes |2\rangle + |\psi_{\rm C}\rangle \otimes |2\rangle = \Big(- |\psi_{\rm B}\rangle + |\psi_{\rm C}\rangle\Big) \otimes |2\rangle,$$

and after passing both beam splitters

$$|\psi_2\rangle = \Big(- |\psi_{\rm D}\rangle + |\psi_{\rm E}\rangle + |\psi_{\rm F}\rangle + |\psi_{\rm G}\rangle\Big) \otimes |2\rangle.$$

The probability density for the atom to have the transverse position q is then given by the square modulus of the wave function $|\psi_2\rangle$ at the given point q. In the left peak (where only the beams D and E contribute) this gives

$$P_{2\,\text{left}}(q) = |\psi_{\rm D}(q)|^2 + |\psi_{\rm E}(q)|^2 - \psi^*_{\rm D}(q)\psi_{\rm E}(q) - \psi^*_{\rm E}(q)\psi_{\rm D}(q).$$

The latter two terms here are responsible for the interference pattern. The internal state is the same in all cases, so it vanishes from the expression.

Consider now the case (b) where the radiation in the microcavities is turned *on*. The state of the atom after passing the first beam splitter is in this case

$$|\psi_1'\rangle = -|\psi_{\rm B}\rangle \otimes |2\rangle + |\psi_{\rm C}\rangle \otimes |3\rangle,$$

and after passing both beam splitters it is

$$|\psi_2'\rangle = -|\psi_{\rm D}\rangle \otimes |2\rangle + |\psi_{\rm E}\rangle \otimes |3\rangle + |\psi_{\rm F}\rangle \otimes |2\rangle + |\psi_{\rm G}\rangle \otimes |3\rangle.$$

The probability density of different positions in the left peak is then

$$P'_{2\,\text{left}}(q) = |\psi_{\rm D}(q)|^2 + |\psi_{\rm E}(q)|^2 - \psi^*_{\rm D}(q)\psi_{\rm E}(q)\langle 2|3\rangle - \psi^*_{\rm E}(q)\psi_{\rm D}(q)\langle 3|2\rangle.$$

Because of the orthogonality of the internal states $|2\rangle$, $|3\rangle$ of the atom, the last two terms disappear, so there is no interference pattern in this case.

This calculation (given in (DNR98)) shows how the which way information influences the interference. It is claimed in the paper (DNR98) that the loss of interference in this case cannot be explained from the point of view of the uncertainty principle. We shall show that this *is* possible although in a more complicated way than for the simple two-slit Which Way experiment of Sect. 3.3.

3.4.3. ANALYSIS OF UNCERTAINTIES IN THE EXPERIMENT

Let us analyse uncertainties arising in the interferometer. To make the analysis simpler we shall consider first the situation in such a way as if there is only one rather narrow beam falling on the interferometer (at the point with the horizontal coordinate x) and then split, first into two and finally into four narrow beams. The incoming beam will be assumed to be sufficiently narrow for its width to be less than the distance between the two beams (formed after splitting) at the moment of passing the second lattice. It is only at the end that we shall take into account many parallel incoming beams so that the whole beam might be wide.

We shall find the uncertainty explanation rather easily in the case where the which way recorder is located around the first beam splitter (as has been described in (DNR98)), but the analysis is more complicated in the modified experiment where the recorder is located at the second lattice.

Recorder at the first lattice
Consider, first, what happens after the particle passes the first beam splitter. The scheme of the beams at this beam splitter is given in Fig. 3.6. If

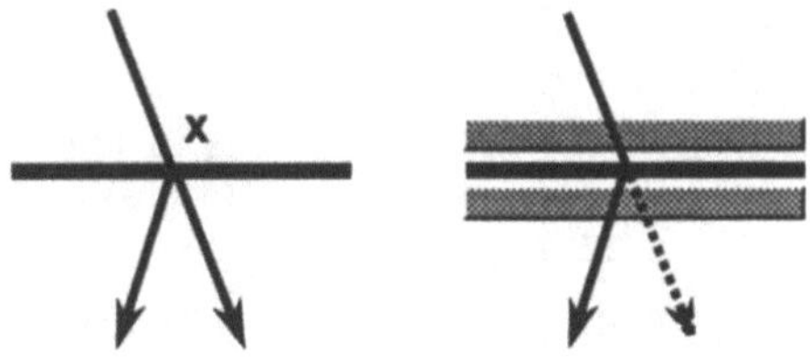

Figure 3.6. The scheme of the first beam splitting: without which way recording (left) and with which way recording (right).

the which way recorder is turned off there are two coherent beams after the beam splitter which may interfere. If it is turned on, the beams after splitting are 'colored' and cannot interfere. One may say that there is only one beam. If the record is read (which is, in principle, possible without violation of the experimental conditions) it is known which of the two beams does exist.

The state of the particle after passing the first beam splitter depends on whether the which way recorder is turned on or off (Fig. 3.7). If it is turned

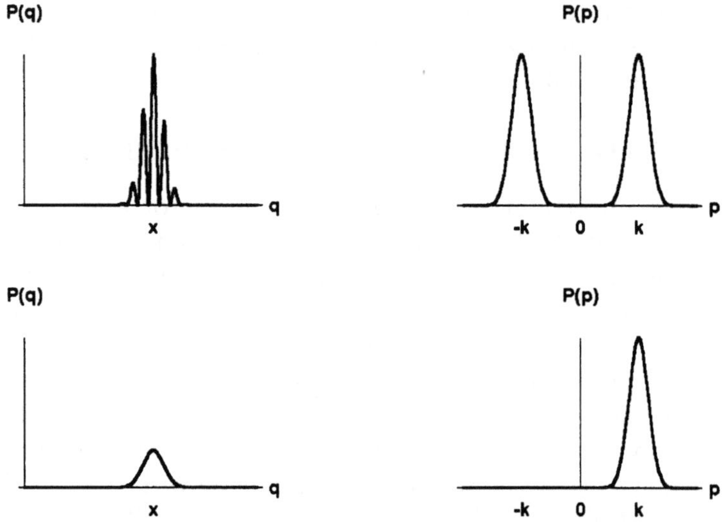

Figure 3.7. The state of a particle after the first beam splitter. Coordinate (left) and momentum (right) representations of the wave function localized near a single point x of the coordinate space but with the linear momentum localized about one value k (lower row) or near two values k and $-k$ (upper row)

off the state (correspondingly in the q- and p-representation) is as drawn in the upper row of diagrams, if the recorder is turned on and the record corresponds to the right beam, the state is as in the lower row of diagrams of Fig. 3.7. In both cases there is only one peak in the q-representation, but in the case of no recording there are two peaks in the p-representation, and therefore the q-peak has periodic structure. The states are quite similar to those in Fig. 3.3 but with q and p interchanged.

This hints how the analysis from the point of view of uncertainties may be carried out for the measurement (which way recording) performed after the first beam splitter (the first lattice). Both from Fig. 3.6 and Fig. 3.7 it is seen that the measurement may be interpreted as a measurement of the linear momentum p with a resolution sufficient to distinguish between the values $p = k$ and $p = -k$ of the momentum. Such a measurement leads to an additional uncertainty of the coordinate q which washes out the periodic structure of the peak in the q-representation (see the upper left and lower left diagrams of Fig. 3.7).

This periodicity has a period equal to the period of the lattice (the beam splitter). The periodicity in the q-representation peak arises because the particle passes through the lattice. If the periodicity is washed out the action of the lattice is cancelled. Therefore the behavior of the particle in the rest of the interferometer may be considered as if there were no first lattice at all. It is evident that the interference pattern will be lost.

We have analysed only the case of a rather narrow beam striking the first lattice in the point x. It is evident that the final conclusion is straightforwardly extended to the case of a wide beam realized as a superposition of many narrow beams with different x. In this case each of the two peaks in p-representation will be narrower and the width of the peak in the q-representation will be wider, but the periodicity inside this wide q-peak will be the same. The state is even simpler because it is close to the superposition of two momentum eigenvalues, with $p = k$ and $p = -k$ (the state may be obtained from that considered in Sect. 3.2.2 by interchanging q and p). If the momentum is measured, only one of two p-peaks will survive and the periodicity of the q-wave function will be lost. Hence the interference pattern will be lost in this case too.

Therefore if the which way information is recorded after the first lattice the loss of the interference may be quite easily explained by the measurement uncertainty relation. In fact only this type of experiment was presented in the paper (DNR98). We can conclude, therefore, that the experiment described in this paper may be explained by the uncertainty relation, contrary to what the authors claimed.

When this explanation was discussed with the authors of (DNR98) they responded that they had performed a simple modification of the experiment which could not be explained in such a simple manner. The modification was that the which way recorder (of the same construction) was located not around the first, but around the second lattice. The interference was lost in this case too.

The uncertainty analysis of this modified experiment is really much more difficult and much more interesting. Let us enter upon it.

The scheme of beams in the general case

The scheme of beams in the interferometer for both locations of the recorder is presented in Fig. 3.8. A single beam is in this case split twice, resulting in four beams after the second beam splitter (a standing wave). The left hand diagram shows the beams if the which way recorder is turned off. The particle is the same in all beams, and the beams can interfere. The middle diagram shows what happens in the case where the which way information is recorded after the first beam splitter (this is the situation analysed above in detail in terms of entanglement). With the recorded information taken into account, there are in fact two different types of particles in the beams. The particles of different types cannot interfere. If the which way recording

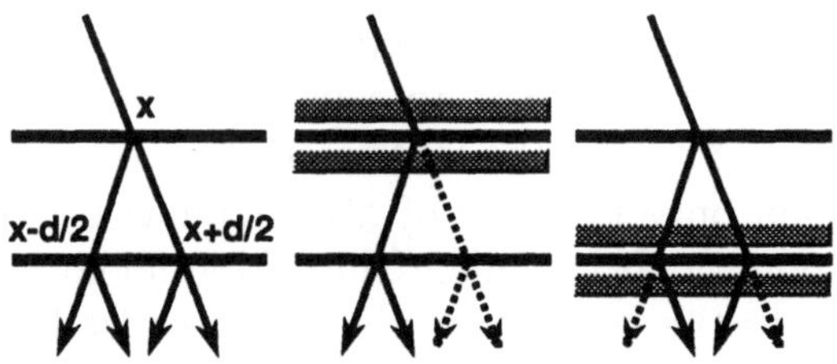

Figure 3.8. Scheme of beams in the interferometer (if there is only one narrow incoming beam): without measurement (left), with the measurement on the first beam splitter (middle) and with the measurement on the second beam splitter (right).

is performed after the second beam splitter (the case only mentioned but not analysed in detail so far) the scheme of beams is presented by the right hand diagram in Fig. 3.8.

To make clearer the final conclusion, let us notice that the beams deflected finally to the right cannot interfere with those deflected to the left. To come to a conclusion about the interference of the beams deflected to the right, we may omit those deflected to the left. Then the the scheme of the beams is finally simplified as in Fig. 3.9. We now see that recording

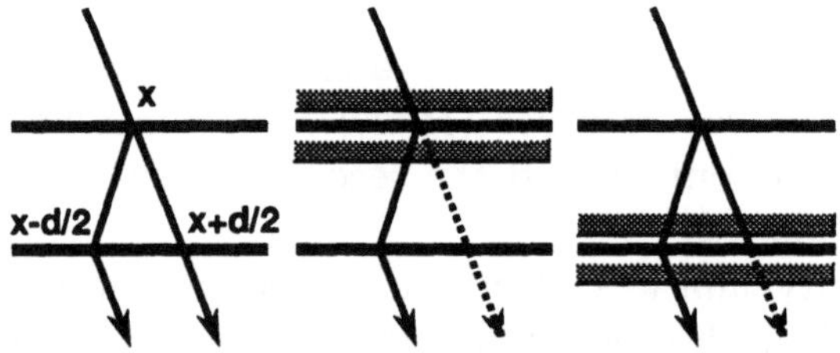

Figure 3.9. Scheme of beams in the interferometer (if there is only one narrow incoming beam): without measurement (left), with the measurement on the first beam splitter (middle) and with the measurement on the second beam splitter (right).

which way information either after the first or after the second lattice is equivalent to a measurement of the position of the particle after the second lattice with the precision better than up to d. The situation is now just as in the two-slit experiment. The interference is owed to the distance d between two locations of the particle after the second lattice. The interference disappears when the particle position is measured that leads to the uncertainty of momentum.

The particle states

Let us illustrate these arguments by the corresponding wave functions. The states of the particle after passing the second beam splitter are presented in Fig. 3.10. The richest structure exists, of course, in the absence of

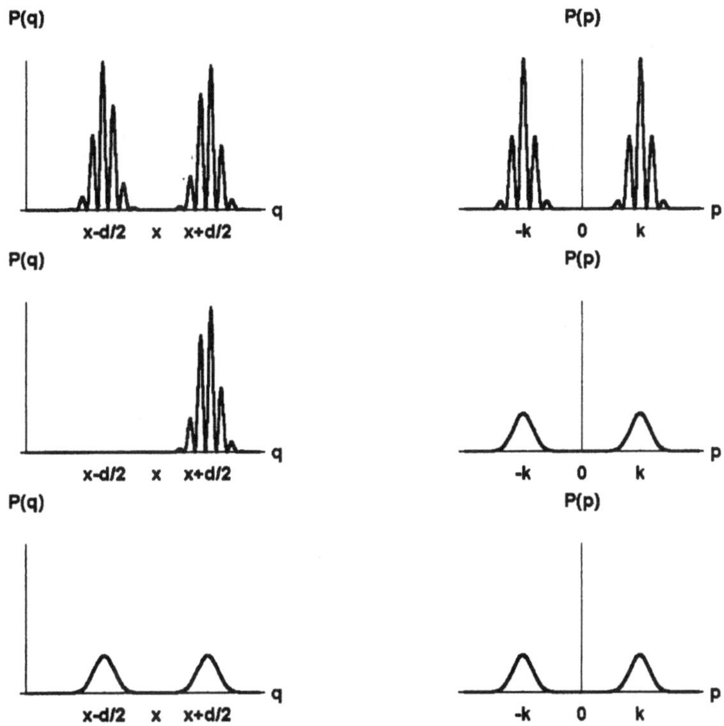

Figure 3.10. Coordinate (left hand) and momentum (right hand) representations of the state after the second lattice in the interferometer: without a measurement (upper row), with the measurement at the first lattice (middle row), and with the measurement at the second lattice (lower row)

any recording (measurement). This is drawn in the upper row of diagrams. There are two peaks in the q-representation corresponding to the alternative locations of the particle at the second beam splitter. Each of these peaks has internal structure because the beam located at a given point may have its linear momentum deflected to the right or to the left (the period of this structure corresponds to the period of the lattice serving as a beam splitter). The two peaks in the p-representation have periodic structure too. It will be discovered as an interference pattern in the far zone.

The which way recording made at the first beam splitter (the middle diagram of Fig. 3.8) annihilates one of the peaks in the q-representation (in accord with the which way information) but maintains the periodic structure of the other (the middle row of Fig. 3.10). The recording made at the second beam splitter (right hand diagram of Fig. 3.8) preserves both peaks in the q-representation but annihilates their periodic structure. In both cases the peaks in the p representation has no periodic structure. This indicates that the interference pattern in the far zone disappears.

The state of the particle without a which way record is the following (x denotes the location of the particle at the first beam splitter):

$$\psi_x(q) \sim \left[e^{-\frac{1}{4\Lambda^2}\left(q-x-\frac{d}{2}\right)^2} + e^{-\frac{1}{4\Lambda^2}\left(q-x+\frac{d}{2}\right)^2}\right] \cos\frac{k(q-x)}{\hbar}, \tag{3.23}$$

$$\begin{aligned} &\exp\left(\frac{i}{\hbar}px\right)\tilde{\psi}_x(p) \sim \qquad (3.24)\\ &\quad \exp\left[-i\frac{(p+k)d}{2\hbar} - \frac{\Lambda^2(p+k)^2}{\hbar^2}\right] + \exp\left[-i\frac{(p-k)d}{2\hbar} - \frac{\Lambda^2(p-k)^2}{\hbar^2}\right]\\ &+ \exp\left[\ i\frac{(p-k)d}{2\hbar} - \frac{\Lambda^2(p-k)^2}{\hbar^2}\right] + \exp\left[\ i\frac{(p+k)d}{2\hbar} - \frac{\Lambda^2(p+k)^2}{\hbar^2}\right]. \end{aligned}$$

The measurement (recording) on the first beam splitter (if the record corresponds, say, to the right beam) annihilates the second term in Eq. (3.23) and the second pair of terms in Eq. (3.24). This is equivalent to the measurement of the particle position after the second lattice. The measurement at the second beam splitter annihilates the first and third, or alternatively second and fourth, terms in Eq. (3.24).

The formulas will be simpler if we take into account only those components of the wave function which correspond to a particle deflected to the right (as in Fig. 3.9). This choice means that the particle's momentum must be close to k, not $-k$. Taking only these components, we have for the state without which way recording

$$\psi_x^{\text{right}}(q) \sim \frac{1}{2}\left[e^{-\frac{1}{4\Lambda^2}\left(q-x-\frac{d}{2}\right)^2} + e^{-\frac{1}{4\Lambda^2}\left(q-x+\frac{d}{2}\right)^2}\right] e^{\frac{i}{\hbar}k(q-x)}, \tag{3.25}$$

$$\tilde{\psi}_x^{\text{right}}(p) \sim e^{-\frac{\Lambda^2(p-k)^2}{\hbar^2}}\left[e^{-\frac{i}{\hbar}p(x+d/2)+\frac{i}{\hbar}kd/2} + e^{-\frac{i}{\hbar}p(x-d/2)-\frac{i}{\hbar}kd/2}\right]. \tag{3.26}$$

The which way recording selects only one term in each of these two formulas. If the record corresponds to the position of the particle at the point $x+d/2$, then the first term is kept in both formulas. From Eq. (3.25) it is clear that this will be measurement of the coordinate. From Eq. (3.26) it is clear that this leads to the disappearance of the periodic structure in the p-representation, i.e., to the disappearance of the interference pattern in the far zone.

Superposition of slit pairs

Up to now we have assumed that only one quite narrow beam of particles falls upon the first lattice at the point x. In a real experiment there is a superposition of a large number of such beams striking the lattice in different points, as shown in Fig. 3.11. As we have seen, the beam falling

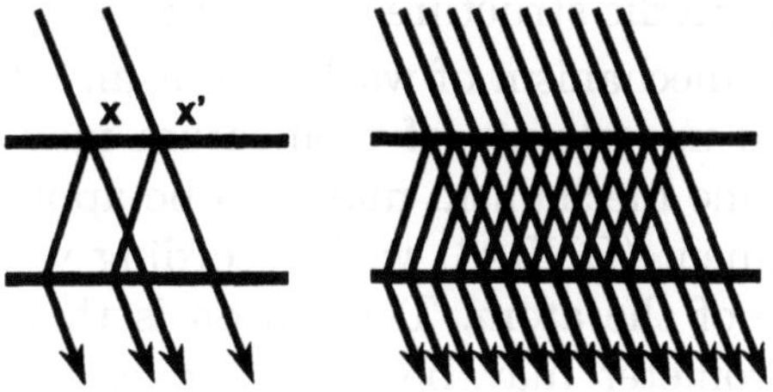

Figure 3.11. Superposition of slit pairs

upon the point x of the first lattice originates the state equivalent to the pair of slits at the points $x \pm d/2$ of the second lattice. The states $\psi_x^{\rm right}$ corresponding to different values of x may be obtained from the universal state (corresponding to $x = 0$) by a shift of the argument:

$$\psi_x^{\rm right}(q) = \psi_0^{\rm right}(q - x), \quad \widetilde{\psi}_x^{\rm right}(p) = \exp\left(-\frac{i}{\hbar}px\right)\widetilde{\psi}_0^{\rm right}(p)$$

If the incoming beam is wide, i.e., is actually described by a superposition of particles striking the first lattice at different points x, then the final state can be evaluated as the corresponding superposition

$$\Psi^{\rm right}(q) = \int c(x)\psi_x^{\rm right}(q)\,dx, \quad \widetilde{\Psi}^{\rm right}(p) = \tilde{c}(p)\widetilde{\psi}_0^{\rm right}(p), \tag{3.27}$$

where we have denoted

$$\tilde{c}(p) = \int c(x)\exp\left(-\frac{i}{\hbar}px\right)\,dx. \tag{3.28}$$

If the function $c(x)$ determining the wide beam is chosen appropriately, the factor $\tilde{c}(p)$ in Eq. (3.27) does not destroy the periodicity (the interference pattern) present in the function $\widetilde{\psi}_0^{\rm right}(p)$. Then the wide beam also demonstrates an interference pattern of approximately the same shape as from a narrow beam.

Thus a wide beam imitates a 'superposition of slit pairs' with approximately the same interference fringes as from a single slit pair. As we have shown earlier, recording which way information in the internal state of the interfering particles is equivalent to the choice of only one slit in each of the superposed pair of slits. Therefore the interference from each slit pair disappears, and the interference of all superposed pairs does too.

Mathematically, if without which way recording the function $\widetilde{\psi}_0^{\text{right}}(p)$ has the form (3.26), then recording leads to the disappearance of one term in the brackets. The periodic structure of this function disappears, and simultaneously the periodic structure of the function $\widetilde{\Psi}^{\text{right}}(p)$ disappears.

Summing up, the mechanism of washing out an interference pattern as a result of additional uncertainty of momentum may be used in the case of a slit pair. The same mechanism may also be applied in the case of the sophisticated experiment (DNR98) with recording which way information in the internal state of the atom. The reason is that: (1) the state of an atom in the interferometer may be presented as a superposition of the states, each corresponding to a pair of slits; (2) the which way recording is equivalent to a measurement of position at each of these slit pairs; and (3) the disappearance of an interference pattern in each of the superposed slit pairs leads to the disappearance of the interference pattern in the whole superposition.

The well known mechanism of the loss of interference because of momentum transfer (expressed by the convolution of uncertainties, see Sect. 3.2) seems to be universal; however, sometimes it is deeply hidden. From the practical point of view the explanation of the loss of an interference pattern by entanglement (Sect. 3.4.2) is, in such situations, much more convenient and 'secure' from the point of view of possible mistakes.

Moreover, the uncertainty relation is applied above in a very special way. We applied the q–p uncertainty relation to each of many components of the particle state, but not to the state as a whole. This is very interesting. The measurement accomplished by the which way recording in the internal state is of an unusual nature. This measurement distinguishes not between different states, but between different histories of the system (this is well illustrated by Fig. 3.8). The above analysis shows that the uncertainty relation has direct relevance to this measurement.

3.4.4. UNCERTAINTY RELATION FOR THE EXPERIMENT

Thus the loss of the interference pattern as a result of 'which way recording' on the second lattice is also connected with 'measurement of the coordinate', but performed in each component of many superposed states (corresponding to narrow beams). The question naturally arises of whether this sophisticated situation may be described formally by a measurement uncertainty relation of the type of those discussed in Sect. 3.1. The problem reduces to finding an appropriate pair of operators A, B.

It is clear that one of these may be the linear momentum operator p, because periodic structure in its distribution gives the interference pattern. We have to find one more operator of such a kind that the information given by the which way record be equivalent to measurement of the second operator. It is clear that the operator of the coordinate q is not appropriate because the particle is not localized in the q-space as a result of which way recording.

The solution, however, becomes possible if we recall that in the present case we consider the particle after it passing the second lattice, and therefore not an arbitrary state from the complete state space of this particle may be realized. Instead, any state of the particle after the second lattice must be of the form

$$\Psi(q) = \int c(x)\psi_x(q)\,dx, \quad \widetilde{\Psi}(p) = \tilde{c}(p)\widetilde{\psi}_0(p), \tag{3.29}$$

with the notations (3.23, 3.24, 3.28). Only functions of this form must be considered as states of the particle. A state of this form may, in fact, be given by the function $c(x)$, because all the other functions in Eq. (3.29) are fixed.

This is why not only operators $\hat{q}$, $\hat{p}$, and functions of them, but also other operators, may be defined in the state space. We need the operator $\hat{x}$ corresponding to the point where the beam falls on the first lattice. This operator may evidently be defined as multiplication of the function $c(x)$ by x:

$$\hat{x}\,\Psi(q) = \int x\,c(x)\psi_x(q).$$

It is easily seen that $\hat{x}$ commutes with $\hat{p}$.

It is evident that the operator

$$\widehat{Q}_1 = \hat{q} - \hat{x} \tag{3.30}$$

is closer to what we need. Being applied to the superposition (3.29), this operator distinguishes between the two components (localized at the points $x+d/2$ and $x-d/2$) of each of the components (3.23) in the superposition. For the first component this operator has a value $d/2$ and for the second component a value of $-d/2$. If this operator is measured (with a resolution better than d), the state (3.29) will change in such a way that only the first (alternatively, only the second) components are left in the superposition.

The measurement of the operator $\widehat{Q}_1$ precisely corresponds (as is readily seen from Fig. 3.8, middle diagram) to the which way recording at the first lattice. The commutation relation

$$[\widehat{Q}_1, \hat{p}] = i\hbar$$

guarantees, owing to the measurement UR (3.14), that the measurement of $\widehat{Q}_1$ with an error $DQ_1 \leq d$ will induce additional uncertainty of $\hat{p}$

$$\delta p \geq \frac{\hbar}{2DQ_1} \geq \frac{\hbar}{2d},$$

which will wash out its periodic structure. This is one more explanation of the loss of an interference pattern in recording at the first lattice.

Forming, now, the operator

$$\widehat{Q}_2 = \frac{1}{k}\hat{p}\,(\hat{q} - \hat{x}), \tag{3.31}$$

we have what we need for the modified experiment. The first factor $\hat{p}/k$ in this operator is approximately equal to ± 1, depending on whether the momentum is equal to k or $-k$. As is seen from Fig. 3.8, right hand diagram, and from Eqs. (3.23, 3.23), this operator has (approximately) the value $d/2$ for those components which correspond to the dashed lines in Fig. 3.8, right hand diagram, and the value $-d/2$ for the solid lines. Therefore the measurement of this operator precisely corresponds to which way recording at the second lattice.

The commutation relation of this operator is

$$[\widehat{Q}_2, \hat{p}] = i\hbar\frac{\hat{p}}{k}.$$

The modulus of the r.h.s. is (approximately) equal to $\hbar$. Therefore according to the general measurement UR (3.14) the measurement of $\widehat{Q}_2$ with the error $DQ_2 \leq d$ will induce an additional uncertainty of $\hat{p}$

$$\delta p \geq \frac{\hbar}{2DQ_2} \geq \frac{\hbar}{2d},$$

which will also wash out its periodic structure.

This is what we needed. The uncertainty relations are proved to be sufficient for explaining the loss of the interference pattern. This does not mean, of course, that they are always necessary. We have seen that the explanation with the aid of entanglement is sometimes much simpler.

3.5. Comments

The uncertainty relations (UR) are (and always have been) one of the most widely discussed aspects of quantum mechanics. They are often discussed in connection with measurements (Boh52; Dir58; Man72; LL87; Mes70; BK92; Ste92). However, usually the original Heisenberg UR for a state are derived in handbooks quite strictly while their application to measurement is only discussed at the qualitative level or illustrated by examples.

In the literature on the mathematically strict theory of quantum measurements the measurement UR are considered in general and precise ways (Hel76; Hol82; Mar91), but in the framework of a complicated mathematical formalism which hides, in fact, the essential physical content of the subject. We have tried to find a compromise between these two approaches and formulate the measurement induced UR in a simple form to emphasise the physical origin of the UR. Limitation of generality may in some cases be even advantageous. What is called by some authors a generic measurement is sometimes a combination of the proper measurement and subsequent transformation of the state. It is clear that relations between the precision of such a 'measurement' and uncertainties after it may be arbitrary. We considered UR only for measurements 'of the first kind' or their direct generalizations.

The question about UR in measurements again became quite actual after the long interesting discussion in the literature of last years (SEW91; Ste92; WH95; ESW95; STCW94; WHC+97; Kni98; DNR98). We have analysed in detail the last and most intriguing experiment discussed in this series of works (DNR98).

Part II

Continuous Quantum Measurements: Theory

Chapter 4

Continuous Quantum Measurements: The General Notion

In the preceding chapters we made the reader familiar with what are usually called instantaneous quantum measurements, although in reality such measurements are not instantaneous but brief. More precisely, the duration of such a measurement is not taken into account. We now go over to continuous quantum measurements. In the present chapter the general notion is given of what a continuous measurement is.

One type of continuous quantum measurement may be most easily imagined as a series of von Neumann's 'instantaneous' reductions. In the present chapter we apply this simple presentation to the analysis of a class of continuous measurements leading to the inhibition of any dynamics, to the freezing of the system in its initial state (the so called quantum Zeno effect). We call such a measurement 'sharp' or 'projective' continuous measurement. It may be also called 'hard' because of its strong influence on the measured system. The presentation of such a measurement given in Sect. 4.1 is taken from the previous book of the author (Men93) with insignificant changes.

The possibility of the quantum Zeno effect induced by a hard measurement is probably the most fascinating illustration of the concept of continuous measurements. However, the opposite case of a soft (fuzzy) continuous measurement is no less interesting. This is because such a measurement not only does not prevent the dynamics of the measured system, but also allows an experimenter (with unavoidable quantum mechanical restrictions) to visualize this dynamics.

We shall discuss this possibility in Chapter 7. Now we illustrate some features of fuzzy (unsharp) or soft continuous measurements by considering the diffusion of a quantum particle through a medium. In this chapter we analyse this type of measurement with the help of explicit physical models, but the subsequent chapters will mostly be devoted to the phenomenological theory of fuzzy continuous measurements.

4.1. Projecting (hard) continuous measurement

As is well known, the classical Zeno paradox says that an arrow cannot fly because time consists of separate instants and an arrow is at a definite point at any instant. In the other words, an arrow being at a definite point at each instant inhibits its flying (motion). Misra and Sudarshan (MS77) gave the name *'quantum Zeno paradox'* to the following statement: frequent (continuous in the limiting case) measurement of a quantum system inhibits its transition into another state. Of course, this statement was derived from quantum mechanics instead of being declared on philosophical grounds, as with the classical Zeno paradox.

The quantum Zeno paradox, or rather, as it is more frequently called quantum Zeno effect, attracted attention and was discussed in a number of theoretical papers, see for example (CSM77; Per80; Sto82; Kha88; NNPR96; Pas97). One more name accepted for this phenomenon in the literature is the *watchdog effect.* We shall give here a short sketch of the theory of the quantum Zeno effect (in the framework of the von Neumann theory) and of its experimental confirmation.

4.1.1. THE THEORY OF THE ZENO EFFECT

Let a system, starting (at the instant $t_0 = 0$) in the state $|\psi_1\rangle$, undergo measurements at the instants $t_1, t_2, \dots$, where $t_i - t_{i-1} = \Delta t$. Let N measurements be performed during the interval T, so that $\Delta t = T/N$. We shall investigate the influence of these measurements on the evolution of the system. Then, letting $N \to \infty$ or $\Delta t \to 0$, we can judge the effect of continuous measurement and its influence on evolution. In this limit the Zeno effect should arise.

The measurement will be described by the set of the two projectors $P_1 = |\psi_1\rangle\langle\psi_1|$ and $P_2 = \mathbf{1} - P_1$. For simplicity one can suggest that the system has only one state $|\psi_2\rangle$ orthogonal to $|\psi_1\rangle$, so that $P_2 = |\psi_2\rangle\langle\psi_2|$. Then each measurement answers the question of which of the two states the system is in. There are two possible outputs of such a measurement, numbered 1 and 2. The output 1 means that the system turns out to be in the state $|\psi_1\rangle$. According to von Neumann's reduction postulate (see Sects. 2.1, 2.3) in this case the state after the measurement is $|\psi_1\rangle$. The other possible output of the measurement is 2. This means that the system turns out to be in the state $|\psi_2\rangle$. And it is actually in this state after the measurement (is reduced into this state).

To find the probabilities of both measurement's outputs one should find the state of the system immediately before the measurement. This state arises in the course of the free evolution of the system between two measurements.

Let the system be in the state $|\psi_1\rangle$ at the instant $t_0 = 0$ and develop freely (according to its own laws without any measurement) during the interval $[t_0, t_1] = [0, \Delta t]$. Let this evolution (governed by the Schrödinger equation) be described by the superposition

$$|\psi(t)\rangle = c_1(t)|\psi_1\rangle + c_2(t)|\psi_2\rangle \tag{4.1}$$

where $c_1(0) = 1$, $c_2(0) = 0$. Then the measurement performed at the instant Δt will give the output 2 with some probability p and the output 1 with the complementary probability $1 - p$:

$$|c_2(\Delta t)|^2 = p, \quad |c_1(\Delta t)|^2 = q = 1 - p.$$

The number p is therefore the probability of the transition $1 \to 2$ during an interval Δt.

If the initial state is $|\psi_2\rangle$ then its evolution is described by the superposition

$$|\psi'(t)\rangle = d_1(t)|\psi_1\rangle + d_2(t)|\psi_2\rangle,$$

with $d_1(0) = 0$, $d_2(0) = 1$. The measurement's results (outputs) 1 and 2 at the moment $t_1 = \Delta t$ are then realized with the probabilities $|d_1(\Delta t)|^2$ and $|d_2(\Delta t)|^2$ correspondingly. These probabilities are expressed by the same numbers as above:[1]

$$|d_1(\Delta t)|^2 = p, \quad |d_2(\Delta t)|^2 = q = 1 - p.$$

The number p turns out to be also the probability of the transition $2 \to 1$ during an interval Δt.

Thus if the state is $|\psi_{i_0}\rangle$ at the instant $t_0 = 0$ then the probability of the measurement output i_1 at the instant $t_1 = \Delta t$ is p for $i_1 \neq i_0$ and $q = 1 - p$ for $i_1 = i_0$. After the measurement is over the system is (owing to reduction) in the state $|\psi_{i_1}\rangle$. Then everything repeats in just the same way with free development during the interval $[t_1, t_2] = [\Delta t, 2\Delta t]$ and measurement at the instant $t_2 = 2\Delta t$. The probability of obtaining at the instant t_2 the same result as at the instant t_1 is $q = 1 - p$, and the probability of obtaining the opposite result is p.

It is evident that everything will repeat again and again. Therefore subsequent measurements will give the outputs i_0, i_1, i_2, ..., i_N with the probability

$$P(i_0, i_1, i_2, \ldots, i_N) = p_1 p_2 \ldots p_N,$$

[1]This is because the orthogonal states $|\psi_1\rangle, |\psi_2\rangle$ evolve, owing to unitary evolution, into the states $|\psi\rangle, |\psi'\rangle$ which also ought to be orthogonal.

where $p_n = p$, if $i_n \neq i_{n-1}$ and $p_n = q = 1 - p$, if $i_n = i_{n-1}$. This product is equal to

$$p^k q^{N-k}$$

if the state is changed k times as a result of measurements and remains unchanged $(N - k)$ times.

A simple calculation shows that the probability for the final state to be the same as the initial state, $i_N = i_0$, (correspondingly opposite, $i_N \neq i_0$) is equal to

$$P\begin{pmatrix} \text{the same} \\ \text{opposite} \end{pmatrix} = \sum_{k_{\text{odd}}^{\text{even}}} \binom{N}{k} p^k q^{N-k} = \frac{1}{2}\left[1 \pm (q-p)^N\right]$$

where the summation is performed over all even (correspondingly odd) numbers between 0 and N.

The first of these two probabilities can be called the probability of the survival of the original state:

$$P_{\text{surv}} = \frac{1}{2}\left[1 + (q-p)^N\right] = \frac{1}{2}\left[1 + (1-2p)^N\right]. \qquad (4.2)$$

Let N measurements be performed during the interval $[t_0, t_N] = [0, T]$ so that $\Delta t = T/N$. Assume that for sufficiently small Δt the probability of spontaneous transition $1 \to 2$ is quadratic in Δt,

$$p = \lambda^2 \Delta t^2 = \left(\frac{\lambda T}{N}\right)^2.$$

Then for large N we have

$$P_{\text{surv}} = \frac{1}{2}\left[1 + \left(1 - 2\frac{(\lambda T)^2}{N^2}\right)^N\right] = \frac{1}{2}\left(1 + e^{-2\frac{\lambda^2 T^2}{N}}\right).$$

In the limit $N \to \infty$, when the measurement can be considered to be continuous, this probability tends to unity,

$$\lim_{N\to\infty} P_{\text{surv}} = 1.$$

This means that continuous observation (measurement) inhibits the transition 1→2, which has a finite probability if no measurement is performed. This is just what is called the *quantum Zeno effect.*

The condition, under which this result was derived, is that the probability of a spontaneous transition $1 \to 2$ during a small time interval Δt is quadratic in this interval, $p \sim \Delta t^2$. For linear time dependence of the probability, $p \sim \Delta t$, no inhibition (no quantum Zeno effect) emerges. This is why this proof of the quantum Zeno effect is invalid for the decay of an unstable state. The point is that the typical law of spontaneous decay is the exponential law

$$p(t) = 1 - e^{-\lambda t}$$

which gives a linear time dependence for time small, $p \sim \Delta t$. This does not mean that there is no Zeno effect for a spontaneous decay, but that this question must be considered in a less formal approach. The question about the Zeno effect in a spontaneous decay has not been solved so far, see, for example, (Pan96; Men99; Pan99). A review on the quantum Zeno effect may be found in (HW97). A new realization of the Zeno effect is discussed in Chapter 9.

4.1.2. EXPERIMENTAL VERIFICATION

Different schemes of experiments for verifying the quantum Zeno effect have been proposed. Among them Cook (Coo88) suggested an experiment with a 3-level system (an ion in a magnetic trap). This scheme was realized later on. The idea of the experiment is the following.

Two levels E_1, E_2 of the system (see Fig. 4.1) play the role of the states $|\psi_1\rangle$, $|\psi_2\rangle$ of Sect. 4.1.1. Transitions between these levels can be induced by radiation having a resonance frequency $\omega_1 = (E_2 - E_1)/\hbar$. The transitions between levels E_1 and E_3 are used for measuring the system's state: the observation of the scattering of radiation in resonance with $\omega_2 = (E_3 - E_1)/\hbar$ allows one to conclude that the system is in the state E_1.

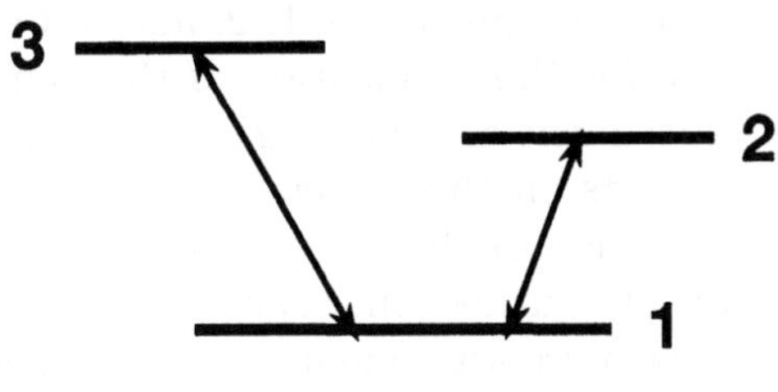

Figure 4.1. A three-level system for verification of the Zeno effect. Repeated observations of whether system is on level 1 or on level 2 makes a transition between them less probable. The level 3 is necessary for the observation.

The probability of spontaneous decay from level 2 to level 1 is assumed to be negligible. The influence of 1–2 resonance radiation leads to the formation of a coherent superposition of the levels 1 and 2 of the type of (4.1) with coefficients c_1, c_2 such that

$$p(t) = |c_2(t)|^2 = \frac{1}{2}(1 - \cos \Omega t) \tag{4.3}$$

Here Ω is the so called Rabi frequency, proportional to the amplitude of the applied resonance radiation. It is important that for small times Δt the probability is quadratic in time:

$$p(\Delta t) = \frac{1}{4}\Omega^2 \Delta t^2,$$

so that the Zeno effect is able to arise.

Transition from level 3 to level 1 is allowed in the proposed experiment, and level 3 can decay only to level 1. If an intense pulse of 1–3 resonance radiation is applied to the system the pulse acts as a measurement corresponding to the projectors P_1, P_2 of Sect. 4.1.1, reducing the system state to one of the states 1 or 2. Indeed, if the system is in the state 1 the transition of it into the state 3 is induced by the action of the resonance radiation (of frequency ω_2), with a consequent spontaneous transition back into the state 1. This is accompanied with absorption and consequent emission of the resonance photon, which is equivalent to scattering of the photon. If the system is in the state 2 no scattering occurs. Thus the observation of scattering of the 1–3 resonance radiation indicates that the system is at level 1, while the absence of scattering means that the system is at level 2.

If the system is in a superposition (4.1) of the states 1 and 2 then one of the following events will happen with the corresponding probabilities $|c_1|^2$, $|c_2|^2$: (1) the resonance photon will be scattered and the system will turn out to be in the state 1; or (2) the photon will not be scattered and the system will be in the state 2 after the pulse has ended.

Thus the pulse of 1–3 resonance radiation acts as a $\{P_1, P_2\}$ measurement, with reduction to one of the states 1, 2, and a 3-level system of this type is quite appropriate for the verification of the Zeno effect. The most advantageous is the experiment as long as $T = \pi/\Omega$. During this time the 1–2 resonance radiation acts on the system (the so called π pulse) inducing the coherent superposition of the states 1 and 2, and n short pulses of 1–3 resonance radiation are turned on, separated by equal time intervals $T/(n+1)$. This means that n measurements are performed during the interval T. If $n = 0$, the probability of th transition $1 \to 2$ is given by Eq. (4.3), and at the time T it is equal to unity, so the system will certainly be in the state 2. However, if $n \neq 0$, the probability of the transition $1 \to 2$ is less than unity, decreasing with an increase of n. For large n this probability is close to zero, expressing the Zeno effect in a clear form.

An experiment close to this scheme has been realized by the group of Wineland (IHBW90). About 5000 ions of Beryllium $^9Be^+$ were stored in a magnetic trap. The levels 1 and 2 were hyperfine sublevels in the ground $2s^2S_{1/2}$ state in a magnetic field. The transition between them was in resonance with the radio frequency 320.7 MHz. The level 3 was one of the sublevels of the $2p^2S_{3/2}$ state, which decays only to level 1. The transition $1 \to 3$ was in resonance with laser radiation of wavelength 313 nm. The 313 nm fluorescence from the ions was detected and served as a signal that some ions were in the state 1.

Before the experiment started the ions had been optically pumped[2] to the ground state 1. Then the 1–2 resonance radio frequency field was applied for as long as $T = 256$ msec, which corresponded to a π pulse. During this radio frequency pulse n short (2.4 msec) 313 nm pulses were applied, where n was 1, 2, 4, 8, 16, 32 or 64. These pulses were long enough to reduce almost all of the ions' wave functions (into one of the states 1 or 2) without causing significant optical pumping.

When the radio frequency π pulse was over, the 313 nm laser radiation was turned on to again prepare the state 1. The number of photons scattered during the first 100 msec of this radiation was counted which was approximately proportional to the number of ions at level 1 at the end of π pulse (the deviation from proportionality was taken into account with the help of calibration).

The measured numbers of ions at level 1 at the instant T were for each n compared with the numbers predicted by the theory. The agreement was almost complete. Thus the theory of the quantum Zeno effect received experimental support. It is interesting that this scheme of experiment precisely reproduces the situation which has been described theoretically by von Neumann's projection (or reduction) postulate.

4.2. Fuzzy (soft) continuous measurement

The quantum Zeno effect takes place when the repeated measurements are sharp (or projective) and are repeated often enough. This is an example of a hard continuous measurement which influences the dynamics of the system very strongly. Let us now consider an opposite example of quantum diffusion when the dynamics is also influenced, but in a softer way.

Monitoring of the position of a (point) particle is a typical continuous measurement. If the particle moves through some medium the states of the atoms of the medium are changed as a result of the interaction with it. As a result, information about positions of the particle at different instants of time is recorded in the state of the medium. This means that the position is monitored. Simultaneously the motion of the particle is influenced by the medium.

[2]This was done with the help of a long (5 sec) 313 nm laser pulse during which those ions that were on level 2 had enough time to decay into level 1 with subsequent repeated induced transitions to level 3 and spontaneous decay into level 1. Finally, when the laser radiation was turned off almost all ions were at level 1.

This rather complicated scheme of measurement may be analysed just as a simple measurement has been in Sect. 2.1: the behavior of both the particle and the medium should be found and then the degrees of freedom of the medium (environment) integrated out. The resulting influence of the medium leads to the decoherence of the particle. This calculation has been made in different ways in the papers (CL83; KMN93; Men97b). The difference between these papers lies in the mechanism for recording the information about the particle in the medium and therefore in the mechanism of the particle's decoherence. We shall now consider these works in very general aspects. In Sects. 4.3, 5.3 two of them will be considered in more detail.

In the first two papers (CL83; KMN93) the following master equation was derived for the density matrix of the measured particle:[3]

$$\dot{\rho} = -\frac{i}{\hbar}[H, \rho] - \frac{1}{2}\kappa[\mathbf{r}\,[\mathbf{r}\,\rho]], \tag{4.4}$$

In (CL83) the information about the position of the particle was assumed to be recorded in the state of the crystal as a whole. Each atom of the crystal was assumed to be a point, the internal motion of atoms being neglected. *Decoherence* is performed in this case *by modes of the crystal.* As a result, the following value has been obtained for the parameter κ:

$$\kappa = \frac{2\eta kT}{\hbar^2} \tag{4.5}$$

where η is the damping constant of the medium, the same as enters the classical equation of motion for the particle in the medium:

$$m\ddot{\mathbf{r}} + \eta\dot{\mathbf{r}} + V'(\mathbf{r}) = F(t) \tag{4.6}$$

This scheme of decoherence may be illustrated by the following diagram:

`Quantum diffusion: Decoherence by crystal modes`

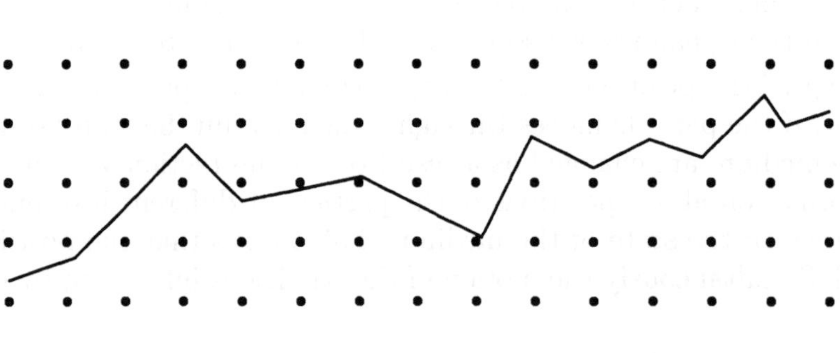

[3]Actually a more general equation was obtained in (CL83), but we consider here only a special case (realized for certain relations between parameters) which is convenient for illustrating the main ideas.

Instead of this, it has been assumed in (KMN93) that the internal state of the atom is changed as a resuslt of its interaction with the bypassing particle. In this case the information about the particle passing an atom may be recorded in its internal state. The resulting master equation is the same as Eq. (4.4) but with

$$\kappa = \frac{2}{l^2 \tau}, \tag{4.7}$$

where l is the distance at which the particle excites an atom and τ a characteristic time of interaction between the atom and the particle. In this case *decoherence* is caused *by the internal structure of atoms.* This may be illustrated by the following diagram:

Quantum diffusion: Decoherence by the structure of atoms

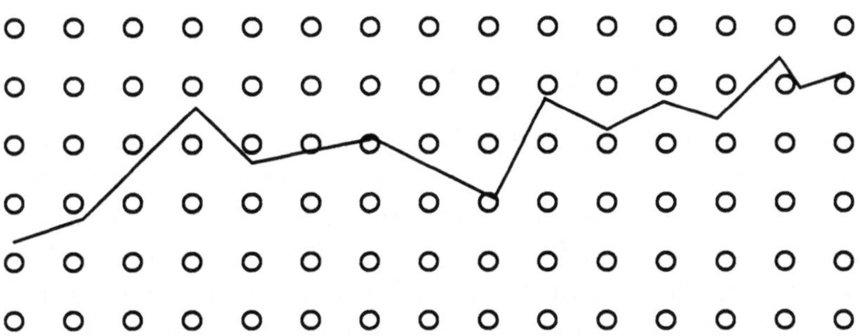

It is clear that in a real situation decoherence by the internal structure of atoms may be stronger (leading to a larger parameter κ) than decoherence by crystal modes.

In the paper (Men97b) one more factor determining the character of decoherence was accounted for, namely a finite resolution of time in the process of continuous measurement (see Sect. 5.3). From the physical point of view this means that the time for establishing equilibrium in the atom is comparable with the other time scales. If finiteness of the resolution cannot be neglected, the evolution of the measured particle cannot be correctly presented by a differential equation for the density matrix.

The master equation (4.4) describes an open continuously measured system. However, this description is non-selective. This means that the information recorded in the environment is not taken into account. Taking this information into account enables one to describe the measured system in a more detailed way (selective description). In principle, selective description may be derived from the models of the type considered. Instead of this we shall obtain it in the framework of the phenomenological restricted-path-integral or complex-Hamiltonian approach (Chapters 5, 6, 7) and return to the model later on (Chapter 8).

4.3. Model of quantum diffusion

Let us consider in more detail the second of the abovementioned models, interpreting it as a model for continuous monitoring of the position of a quantum particle (KMN93).[4] The model of the medium contains a net of quantum oscillators, each having a finite range of interaction with the particle. The positions of the oscillators are fixed, so the model gives a simple presentation of the internal structure of atoms, neglecting the motion of their centers of mass.

Taking the trace over the variables of the measuring oscillators (medium, environment), we shall obtain a density matrix for the particle undergoing influence by the continuous measurement. We shall find both an integral representation of this density matrix and a differential equation for it (the master equation).

4.3.1. THE MODEL OF THE MEASUREMENT

The measuring device to be considered is a system of an infinite number of non-interacting quantum oscillators located at the nodes of a cubic lattice (the three-dimensional lattice will be considered although the generalization onto fewer dimensions is straightforward). The oscillators are labelled by the index k ($k = 1, 2, \ldots$). They are in the ground state before the measurement. The range of interaction of an oscillator with the particle is finite, hence its excitation indicates that the particle passes in its vicinity. The interaction of the kth oscillator with the measured particle is described by the following Hamiltonian:

$$H_{\text{int}} = \sum_k H_k = \sum_k \gamma q_k \exp\left[-\frac{(\mathbf{r}-\mathbf{c}_k)^2}{l^2}\right]. \tag{4.8}$$

Here q_k is a canonical coordinate of the oscillator while $\mathbf{c}_k$ is its location, γ is the interaction constant, l characterizes its range, and $\mathbf{r}$ is the coordinate of the particle measured.

It should be emphasised that the canonical coordinate q_k of the kth oscillator has nothing to do with its location in space. It is only under this condition that the oscillator can measure the coordinate $\mathbf{r}$ of the particle by registering its passing the point $\mathbf{c}_k$ at which the oscillator is located. In the case of a two- or one-dimensional lattice q_k may denote the displacement in the spacial direction orthogonal to those dimensions in which

[4]This section is reprinted, with appropriate changes, from "Physics Letters A", Volume 177, A.Konetchnyi, M.Mensky and V.Namiot, The physical model for monitoring position of a quantum particle, Pages 283–289, Copyright 1993, with the permission of Elsevier Science

the particle moves. In the three-dimensional case oscillators may have an electromagnetic or some other non-mechanical nature.

The complete Hamiltonian of the problem has the form

$$H = \frac{\mathbf{P}^2}{2M} + V(\mathbf{r}, t) + \sum_k \frac{p_k^2}{2m} + \sum_k \frac{m\omega^2 q_k^2}{2} + H_{\mathrm{int}}. \tag{4.9}$$

Here $\mathbf{P}$, M are the linear momentum and the mass of the measured particle.

The density matrix for the complete system which consists of the measured particle and the net of oscillators may be expressed in the form of a Feynman path integral. Integrating then over the coordinates q_k of all the oscillators one has the density matrix for the particle. As a result we find the density matrix for the measured particle in the form

$$\begin{aligned} \rho_T(\mathbf{r}_1, \mathbf{r}_2) &= \int \mathrm{d}\mathbf{r}_{01} \int \mathrm{d}\mathbf{r}_{02}\, U(\mathbf{r}_1, \mathbf{r}_{01} | \mathbf{r}_2, \mathbf{r}_{02}) \rho_0(\mathbf{r}_{01}, \mathbf{r}_{02}), \\ U_T(\mathbf{r}_1, \mathbf{r}_2 | \mathbf{r}_{01}, \mathbf{r}_{02}) &= \int_{\mathbf{r}_{01}}^{\mathbf{r}_1} \mathrm{d}[\mathbf{r}_1] \int_{\mathbf{r}_{02}}^{\mathbf{r}_2} \mathrm{d}[\mathbf{r}_2] \exp\left\{ \frac{iM}{2\hbar} \int_0^T \mathrm{d}t \left(\dot{\mathbf{r}}_2^2 - \dot{\mathbf{r}}_1^2 \right) \right. \\ &- \left. \frac{i}{\hbar} \int_0^T \mathrm{d}t \left[V(\mathbf{r}_2, t) - V(\mathbf{r}_1, t) \right] \right\} F[\mathbf{r}_1, \mathbf{r}_2]. \end{aligned} \tag{4.10}$$

Here $F[\mathbf{r}_1, \mathbf{r}_2]$ is the influence functional (Fey48; FV63; FH65) of the oscillators (measuring system) on the particle (the measured system). It may be found from the solution to the Schrödinger equation for the oscillators with the given external force:

$$i\hbar \frac{\partial \psi}{\partial t} = -\frac{\hbar^2}{2m} \frac{\partial^2 \psi}{\partial q^2} + \frac{m\omega^2 q^2}{2} \psi + \gamma q f(t) \psi.$$

For the ground state as the initial condition

$$\psi(q, 0) = \mathcal{N} \exp\left(-\frac{m\omega q^2}{2\hbar} \right)$$

the solution is

$$\psi(q, t) = \mathcal{N} \exp\left(-\frac{m\omega q^2}{2\hbar} - \frac{m\omega q s(t)}{\hbar} - \frac{i\omega t}{2} + \frac{im\omega^2}{2\hbar} \int_0^t s^2(t) \mathrm{d}t \right), \tag{4.11}$$

where

$$s(t) = \mathrm{e}^{-i\omega t} \frac{i\gamma}{m\omega} \int_0^t \mathrm{e}^{i\omega t} f(t) \mathrm{d}t.$$

The influence functional is expressed through the wave function as follows:

$$F[\mathbf{r}_1, \mathbf{r}_2] = \prod_k \int_{-\infty}^{\infty} dq_k \psi^*_{k[\mathbf{r}_2]}(q_k) \psi_{k[\mathbf{r}_1]}(q_k). \tag{4.12}$$

Here $\psi_{k[\mathbf{r}]}(q)$ is the solution of the form (4.11) for the driving force

$$f_{k[\mathbf{r}]}(t) = \exp\left(-\frac{(\mathbf{r}(t) - \mathbf{c}_k)^2}{l^2}\right). \tag{4.13}$$

The form of the wave function $\psi_{k[\mathbf{r}]}(q)$ depends on the location $\mathbf{c}_k$ of the kth oscillator. In addition to this it is a functional of the trajectory $[\mathbf{r}]$ of the measured particle.

4.3.2. CALCULATION

All integrals in the expression (4.12) for the influence functional are of Gaussian type. They may be calculated explicitly. This gives

$$\begin{aligned} F[\mathbf{r}_1, \mathbf{r}_2] &= \exp\Big\{ -\frac{\gamma^2}{4\hbar m\omega} \sum_k \int_0^T dt \int_0^T dt' \, e^{i\omega(t-t')} \Big[f_{k[\mathbf{r}_1]}(t) f_{k[\mathbf{r}_1]}(t') \\ &\quad + f_{k[\mathbf{r}_2]}(t) f_{k[\mathbf{r}_2]}(t') - 2 f_{k[\mathbf{r}_1]}(t) f_{k[\mathbf{r}_2]}(t') \Big] \Big\}, \end{aligned} \tag{4.14}$$

where the notation (4.13) is used.

To exclude dependence on concrete locations $\mathbf{c}_k$ of the oscillators let us assume that they are distributed homogeneously in space, and then let us go over to a continuous distribution with density n_l. Then the sum in (4.14) becomes an integral which can be calculated. It is convenient (although not necessary) to choose the density n_l in such a way that one oscillator is in a volume of the order of l^3 (l being the distance at which an oscillator feels a particle). Let

$$n_l \left(\frac{\pi l^2}{2}\right)^{3/2} = 1.$$

Then Eq. (4.14) takes the form

$$\begin{aligned} &F[\mathbf{r}_1, \mathbf{r}_2] = \\ &\exp\left\{ -\frac{\gamma^2}{4\hbar m\omega} \int_0^T dt \int_0^T dt' \, \cos\omega(t-t') \left[\exp\left(-\frac{(\mathbf{r}_1(t) - \mathbf{r}_1(t'))^2}{2l^2}\right) \right.\right. \\ &\left.\left. + \exp\left(-\frac{(\mathbf{r}_2(t) - \mathbf{r}_2(t'))^2}{2l^2}\right) - 2\exp\left(-\frac{(\mathbf{r}_2(t) - \mathbf{r}_1(t'))^2}{2l^2}\right) \right] \right\}. \end{aligned} \tag{4.15}$$

4.3.3. IMPROVEMENT OF THE MODEL: TIME IS MEASURED

It is seen from the structure of the influence functional (4.16) that the state of the particle at the instant T depends on the whole process of its preceding evolution, thus the density matrix (4.10) cannot be described by a differential equation. The dependence of the state on the preceding evolution may be removed if the type of the measurement is changed in such a way that the measuring device determines not only the coordinate of the particle but also the instant of time when it was at that point.

Let us assume that measuring oscillators are marked by two indices k and i and have the following interaction Hamiltonians:

$$H_{ki} = \gamma q_k \exp\left[-\frac{(\mathbf{r}-\mathbf{c}_k)^2}{l^2} - \frac{(t-\theta_i)^2}{\tau^2}\right]. \tag{4.16}$$

This means that the oscillator with the number (k, i) is located at the point $\mathbf{c}_k$, is turned on at the instant θ_i, and interacts with the particle during a time interval of the order of $\tau/\sqrt{2}$.

Since the oscillators do not interact with each other this change leads to a simple modification of the influence functional (4.16), which should now depend on the moments θ_i of the oscillators turning on. Just as with the spacial distribution, we may go over to a continuous distribution in time. Let (as a matter of convenience) the time density n_τ of the oscillators be such that one of them lies in the period of the order of τ:

$$\tau n_\tau \sqrt{\frac{\pi}{2}} = 1.$$

In addition to this, assuming that $\omega T \ll 1$, we may set $\cos\omega(t-t') = 1$ in Eq. (4.16). This corresponds to the assumption that the oscillators absorb energy by small quanta, and typical times for the problem are not too large in comparison with the period of the oscillators. The approximations mentioned give the following formula for the influence functional:

$$\begin{aligned} F_T[\mathbf{r}_1, \mathbf{r}_2] &= \exp\left\{-\frac{\gamma^2}{4\hbar m\omega}\int_0^T \mathrm{d}t \int_0^T \mathrm{d}t' \, \exp\left(-\frac{(t-t')^2}{2\tau^2}\right)\right. \\ &\times \left[\exp\left(-\frac{(\mathbf{r}_1(t)-\mathbf{r}_1(t'))^2}{2l^2}\right) + \exp\left(-\frac{(\mathbf{r}_2(t)-\mathbf{r}_2(t'))^2}{2l^2}\right)\right. \\ &\left.\left. -2\,\exp\left(-\frac{(\mathbf{r}_2(t)-\mathbf{r}_1(t'))^2}{2l^2}\right)\right]\right\}. \end{aligned} \tag{4.17}$$

As a result of measuring the time, the trajectories $[\mathbf{r}_1]$, $[\mathbf{r}_2]$ in the functional (4.17) are compared at the instants differing by not more than τ. To guarantee effective interaction of the oscillators with the measured particle let us consider particles of not very large energy,

$$E_0 \ll \frac{Ml^2}{2\tau^2},$$

which corresponds to a sufficiently long time of flight of the particle within the limits of the range of interaction. Under this condition the value of the functional (4.17) is exponentially small for trajectories differing by more than l at close instants (coinciding up to τ). Therefore the exponent of (4.17) may be calculated in the lowest order in $\mathbf{r}_{12} = \mathbf{r}_2 - \mathbf{r}_1$. The exponent being negatively determined, the expansion in $\mathbf{r}_{12}$ begins with the second degree term. Furthemore, owing to the assumption about relatively small initial energy of the particle we may neglect the contribution of trajectories corresponding to velocities greater than l/τ. All this gives the following formula for the influence functional in the case of measured time:

$$F_T[\mathbf{r}_1, \mathbf{r}_2] = \exp\left(-\frac{\gamma^2\tau}{4m\hbar\omega l^2}\sqrt{2\pi}\int_0^T (\mathbf{r}_2(t) - \mathbf{r}_1(t))^2\, dt\right).$$

Let us accept the following restriction on the parameters of the measuring device:

$$\frac{\gamma\tau}{m\omega} = \nu\sqrt{\frac{\hbar}{m\omega}},$$

where ν is a number greater than or of the order of unity. This condition means that the shift of the oscillator coordinate owed to the influence of the particle is greater than the width $\sqrt{\hbar/m\omega}$ of the ground state of the oscillator. For convenience take $\nu^2 = 4/\sqrt{2\pi}$. Then

$$F_T[\mathbf{r}_1, \mathbf{r}_2] = \exp\left[-\frac{1}{l^2\tau}\int_0^T (\mathbf{r}_2(t) - \mathbf{r}_1(t))^2\, dt\right]. \tag{4.18}$$

Substituting the expression (4.18) for the factor F in Eq. (4.10) we easily see that the density matrix satisfies the following *master equation:*

$$\frac{\partial\rho}{\partial T} = -\frac{i}{\hbar}[H, \rho] - \frac{1}{l^2\tau}(\mathbf{r}_2 - \mathbf{r}_1)^2\rho. \tag{4.19}$$

This equation coincides with Eq. (4.4) because of the relation

$$\langle\mathbf{r}_2|\,[\mathbf{r}[\mathbf{r}\rho]]\,|\mathbf{r}_1\rangle = (\mathbf{r}_2 - \mathbf{r}_1)^2\,\langle\mathbf{r}_2|\rho|\mathbf{r}_1\rangle.$$

Notice that in this consideration we have, in fact, replaced the exponential $\exp\left(-(t-t')^2/2\tau^2\right)$ in Eq. (4.17) by the delta-function. This means that the precision of measuring the time is assumed to be absolute. This is possible when the relaxation time of the atoms can be neglected. The resulting dynamics is Markovian and may be described by the master equation (4.19). Most of the considerations in the next chapters will be performed in this approximation. However, in Sect. 5.3 the same process of quantum diffusion will be considered in a more precise approach which takes into account the relaxation time of atoms leading to the final precision of measuring the time.

4.3.4. HEATING OF THE MEASURED PARTICLE

If H is the Hamiltonian of the free particle the equation (4.19) implies the following formulas for mathematical expectations:

$$\frac{\mathrm{d}\langle x\rangle}{\mathrm{d}t} = \frac{1}{M}\langle p\rangle, \qquad \frac{\mathrm{d}\langle p\rangle}{\mathrm{d}t} = 0, \qquad \frac{1}{2M}\frac{\mathrm{d}\langle p^2\rangle}{\mathrm{d}t} = \frac{\hbar^2}{\kappa M}. \tag{4.20}$$

The last equation means that the dispersion of the momentum increases linearly. This is in complete correspondence with the property that the error of the measurement Δa_T decreases in a continuous measurement inversely proportional to $\sqrt{T}$:

$$\Delta a_T = \frac{1}{\sqrt{\kappa T}} = l\sqrt{\frac{\tau}{2T}}$$

(see Chapter 7). Then, in correspondence with the uncertainty principle, the uncertainty of the linear momentum should be proportional to $\sqrt{T}$. This is confirmed by the last equation of Eq. (4.20). One may easily see that the product of both uncertainties is a constant of the order of $\hbar$.

The energy of the free particle is proportional to the square of its linear momentum. Therefore the energy of the particle also increases linearly, with the characteristic time scale

$$T_1 = \frac{2E_0 M l^2 \tau}{\hbar^2},$$

where E_0 is the initial value of the average kinetic energy.

It is necessary to note that as a consequence of a remark made earlier (in Sect. 4.3.3, after Eq. (4.17)) it is senseless to consider the present problem at too large time scales for which $\langle v\rangle$ becomes more than l/τ. Therefore T should be less than $Ml^4/\tau\hbar^2$.

4.4. Comments

A direct and physically transparent analysis of continuous measurements with the help of explicit models of continuous measurements goes back to works of Joos and Zeh (JZ85) for quantum mechanical systems and of Milburn with collaborators (WCM85; Mil87; GWM93; WM94) for systems in quantum optics. Many interesting and realistic mechanical models of quantum measurements are presented in the book of Braginsky and Khalili (BK92). An original technique of quantum trajectories for the consideration of models of continuous measurements has been elaborated by Carmichael (Car93). An excellent work on the behavior of a particle continuously monitored by the medium was produced by Caldeira and Leggett (CL83), although they aimed at analysing quantum diffusion and never mentioned continuous measurement. In the paper (KMN93) the role of internal degrees of freedom of atoms in the medium was analysed in the same physical process, and in (Men97b) the finite relaxation time of atoms was taken into account.

Chapter 5

Fuzzy Continuous Measurements: Phenomenology

It was shown in Chapter 2 that the situation of quantum measurement may be correctly represented by a model including the system measured and its environment (the measuring system or the measuring medium). However, this rather cumbersome procedure of first including the environment into the consideration and then excluding it turns out to be unnecessary. One may consider explicitly only the measured system, while the influence of the measurement (i.e., the influence of the interaction with the environment) may be taken into account implicitly. This means that the system in question is treated as an open one, but the influence of its environment (a continuous measurement) is presented phenomenologically.

For an instantaneous measurement an analogous phenomenological description is supplied by the von Neumann projection and its various generalizations (Sect. 2.3). We now begin to discuss the phenomenological description of continuous quantum measurements (we shall see that this is also a phenomenological description of decoherence, see Sect. 7.2). Of course, this is much more complicated than von Neumann's projection, but the idea is the same. The difference is that we now project quantum processes instead of projecting quantum states.

There are various approaches which realize this program. The conceptually simplest one identifies (at least approximately) a continuous measurement with a sequence of instantaneous (or brief) fuzzy measurements. Representing the instantaneous measurements by operators generalizing von Neumann's projectors, while evolution between them is described by unitary evolution operators, we may obtain a continuous measurement as a product of these. The resulting conceptually simplest, but technically inconvenient, way is discussed in Sect. 5.4 together with some other phenomenological descriptions (the master equation, the stochastic Schrödinger equation, consistent histories) of continuous measurements.

The main part of the present chapter is devoted to another phenomenological approach to continuous measurements which uses restricted path integrals (RPI). On the basis of the Feynman theory of amplitudes, the theory of open continuously measured systems is formulated by restricting

Feynman path integrals to subsets of paths. The final mathematical machinery is reduced to complex Hamiltonians and the corresponding effective Schrödinger equations. This way is followed in the most of the present chapter. After a general formulation in Sect. 5.1 of the selective description of a continuous measurement (when the measurement readout is taken into account) we derive in Sect. 5.2 a master equation giving its non-selective description and give (Sect. 5.3) an example of the measurement which is 'integral in time', i.e. it cannot be represented as gradually evolving in time, step-by-step. Finally, a review of other phenomenological approaches to continuous measurements and the relations of them to the RPI approach is given in Sect. 5.4).

The restricted-path-integral (RPI) approach to continuously measured systems (also called the theory of quantum corridors) plays a very important role in the conceptual structure of the theory. Its conceptual aspect will be discussed in greater detail in Chapter 6. The complex-Hamiltonian approach is technically simpler and therefore is important for applications (Chapters 7, 8).

5.1. Continuous measurements from restricted path integrals

The representation of continuous measurements by restricted path integrals (RPI) was proposed by Feynman in 1948 (Fey48) and technically developed by the present author in 1979 (Men79a; Men79b) and in later papers and books. This approach has conceptual advantages: it is phenomenological (needs no model of the measurement), it is derivable from first principles, and it underlines the role of information in the dynamics of the measured system.

For practical aims advantageous is the method of complex Hamiltonians which is equivalent to RPI in some special but very wide and important class of continuous measurements, monitoring with absolute time resolution.

5.1.1. THE FEYNMAN THEORY OF AMPLITUDES

The concept of an amplitude or, more correctly, probability amplitude has been one of the basic concepts in quantum mechanics from the very beginning. It is generally believed that the main new concept in quantum mechanics (as compared with classical mechanics) is operators instead of c-number observables. However, Dirac argued (Dir72) that the most important distinction of quantum theory is not in operators but in amplitudes. And this is reasonable, because the operator formulation of quantum mechanics is rather abstract, and bacause it is only in terms of amplitudes that the physical content of the theory is completely revealed.

The deepest formulation and most efficient applications of the theory of amplitudes has been given by Richard Feynman in connection with his path-integral approach to quantum mechanics (FH65). It was well known, even before Feynman, that the probability may be obtained as a square modulus of an amplitude, $P = |A|^2$. Feynman clearly formulated that in some physical conditions only amplitudes make sense, and in some other conditions the corresponding probabilities are also meaningful.

The difference between these situations is very simple. If some event may occur through one of several alternative channels characterized by the amplitudes $\{A_i\}$ then for the description of the event one must: (i) sum these amplitudes, $A = \sum A_i$, if there is no way of finding out which channel is actually realized; or (ii) sum the probabilities, $P = \sum P_i = \sum |A_i|^2$, if there is a way of identifying the channel.

In the second case we have, in fact, the situation of measurement. It is realized by certain material objects playing the role of measuring apparatus or a measuring medium. The measured system is then considered as an open system.

This may be easily illustrated (FH65; FLS65) by the modified two-slit experiment, a variant of the Which Way experiment (see Sect. 3.3 and Fig. 3.2). If a flux of particles falls on the film after passing through two slits in an opaque screen the film then shows an interfering pattern. This is a classical two-slit experiment. It is explained by the rule of adding amplitudes. The probability amplitude A to arrive at a certain point on the film is a *sum of the amplitudes* A_1, A_2 to arrive there through the corresponding slits. The probability $P = |A|^2$ for the particle to arrive to the point is therefore

$$P = |A_1|^2 + |A_2|^2 + A_1 A_2^* + A_1^* A_2 \tag{5.1}$$

The two last terms in the sum are responsible for the emergence of the interference pattern and are therefore called 'interference terms'.

However, according to Feynman the rule of adding amplitudes is valid only if there is, in principle, no possibility of finding out which of the two slits the particle passed through. If there is some way of doing this (for example, a flux of photons falls on the slits in such a way that some of photons are scattered by our particle near one or another of the slits) then the complete probability of the particle to reach a certain point has to be evaluated by *summing up the probabilities*:

$$P = P_1 + P_2 = |A_1|^2 + |A_2|^2 \tag{5.2}$$

There are no interference terms (and no interference) in this case.

The latter situation is, in fact, a situation of measurement: the path of the particle is measured by scattering photons. In this case the flux of photons is a measuring device. The probabilities may be ascribed to separate channels leading to the event (certain slits in our case) only because the measurement is performed. However, the measuring device and its interaction with the particle is not included in the calculation Eq. (5.2) explicitly. Instead, the system measured (the particle) is considered as an open system while the influence of the measurement is taken into account implicitly.

This is what we are doing when considering continuous measurements in the framework of the Restricted-Path-Integral (RPI) approach.

5.1.2. RESTRICTED PATH INTEGRALS (RPI)

The idea of the RPI approach to open continuously measured systems has been elaborated by the author (following a short remark of Feynman (Fey48)) in a long series of papers beginning with (Men79a; Men79b) (see also (Kha81; BLP82; Cav86)).[1] The main idea in its modern formulation is that the Feynman integral for the propagator (i.e., for the evolution operator) of a closed quantum system has to be modified in the case of an open continuously measured system according to the information given by the measurement.

Integration in the Feynman path integral for a *closed* (absolutely isolated) system must be performed over all paths connecting the initial and final points because no information could possibly exist in this case about the path chosen by the system (one may say that the system moves simultaneously through all paths). Therefore the propagator (or the evolution operator) of the system may be expressed as an integral over all paths connecting the initial and final points (see Fig. 5.1 a). This gives the Feynman

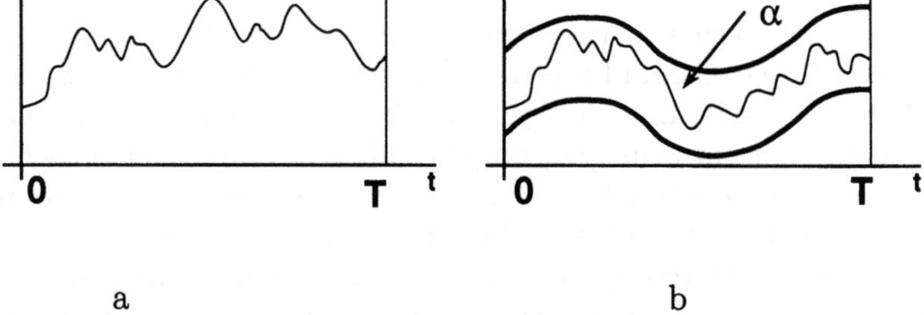

Figure 5.1. The Feynman path integral over all paths (diagram a) and restricted path integrals over paths belonging to the corridor α (diagram b).

[1]There were also unsuccessful attempts to realize Feynman's idea in (BB74; Fri76). The mistake of these authors was to use repulsive real potentials at the border of corridors instead of 'absorbing' imaginary potentials (see Sect. 5.1.3).

path integral which can be written in the coordinate representation[2] as

$$U_T(q'', q') = \int d[q]\, e^{\frac{i}{\hbar} S[q]}, \tag{5.3}$$

or in the phase-space representation as

$$U_T(q'', q') = \int d[p]d[q]\, e^{\frac{i}{\hbar} \int_0^T (p\dot{q} - H(p,q,t))} \tag{5.4}$$

The evolution of the system (in terms of the wave function or the density matrix) is described by the evolution operator

$$|\psi_t\rangle = U_t|\psi_0\rangle, \quad \rho_t = U_t \rho_0 U_t^\dagger,$$

which has values of the propagator as its matrix elements:

$$U_t(q'', q') = \langle q''|U_t|q'\rangle.$$

If the system is *continuously measured* and the measurement results in the readout α, the path integration has to be restricted to the subset of the paths compatible with this readout (see Fig. 5.1 b). The restriction may be described by a *weight functional* $w_\alpha[p, q]$, which must be close to unity for the paths $[p, q]$ compatible with α, and close to zero otherwise. Therefore the path integral takes the form

$$U_T^\alpha(q'', q') = \int d[p]d[q]\, w_\alpha[p, q]\, e^{\frac{i}{\hbar} \int_0^T (p\dot{q} - H(p,q,t))}, \tag{5.5}$$

while the evolution of the system is described as follows

$$|\psi_T^\alpha\rangle = U_T^\alpha|\psi_0\rangle, \quad \rho_T^\alpha = U_T^\alpha \rho_0 \,(U_T^\alpha)^\dagger\,. \tag{5.6}$$

The evolution of a continuously measured (open) system is described, therefore, not by a single evolution operator but by a set of *partial evolution operators* U_T^α with the corresponding *partial propagators.*

The logic followed above led us to the path integral restricted to a set of paths corresponding to the measurement readout α. This is an analogue of the von Neumann projection. As in the von Neumann case the projection depends on the measurement's result (more precisely, on the information supplied by the measurement), but now it is a projection onto a set of paths rather than a projection onto a subspace of states. This is quite natural, of course, because now we deal with continuous measurements which give

[2]The definition of the path integral may be found in Appendix A3, but it may be not necessary in many cases because, technically, path integrals may be reduced to solutions of the Schrödinger equation.

information about the system's evolution, not about the state of the system. The set of paths corresponding to the measurement's readout will be called a *quantum corridor.*

In fact it is the quantum corridor which presents the measurement readout quite adequately because it adequately reflects the information supplied by the readout. For example monitoring of the coordinate gives the readout presented by a curve $[a] = \{a(t)\}$, but the measurement's resolution Δa is very important too. The corresponding quantum corridor will include all curves which differ from $[a]$ by not more than Δa. Thus it takes into account both the curve $[a]$ and the resolution Δa.

Technically, the restriction of the path integral may be described by a weight functional in the integrand, $w_\alpha[q]$ in the coordinate representation of the path integral (this is applicable not for all measurements), or $w_\alpha[p,q]$ in the phase space representation (always applicable). Restriction of the path integral to a certain set of paths is described by the functional w_α equal to unity for the paths belonging to this set and zero otherwise. This is a direct analogue of the von Neumann projection.

We can generalize the definition by assuming that the functional is smooth. It must then be equal to unity for some narrow set of paths 'in the middle' of the quantum corridor and has to gradually decrease for the paths further and further from this set, becoming negligible at the edge of the corridor and outside it. In this case there is no set of paths which can be called a quantum corridor. Instead the quantum corridor, in a somewhat generalized sense, is represented by the given smooth functional $w_\alpha[p,q]$.

If the weight functional w_α is positive (the natural generalization of the characteristic functional equal to either zero or unity) we have an analogy of the fuzzy measurement of Sect. 2.3 with the transformation $|\psi\rangle \to R_a|\psi\rangle$ by positive operators R_a. The corresponding continuous measurement is 'minimally disturbing'. Generalizing further, we can describe a non-minimally disturbing continuous measurement by the complex-valued functional w_α. Yet the case of positive weight functionals is the most important, and we shall assume this property in most applications to be considered.

Each partial evolution operator cannot be unitary because it describes only one channel of evolution. The unity probability is distributed in this case into various channels, and it is only the sum over all channels that corresponds to the total (unity) probability. The precise formulation of this is that the probability density of the readout α is equal to

$$p(\alpha) = \mathrm{tr}\left(U_T^\alpha \rho_0 \,(U_T^\alpha)^\dagger\right). \tag{5.7}$$

Eq. (5.7) is meaningful only if the normalization of the partial propagators is fixed somehow. This may be done naturally by consideration of the non-selective description of the measurements when the concrete measurement readout is not known. In this case all possible measurement readouts must be taken into account, so the summation (actually, integration) must be carried out over the set of alternative measurement readouts α. The summation of density matrices (probabilities) has to be made rather than of state vectors (amplitudes). This follows from the general rules of dealing with amplitudes (Sect. 5.1.1) and the fact that it is, in principle, possible to know which measurement readout is realized (the information about this is recorded in some form or another in the environment of the measured system).

Thus the non-selective description of the same physical situation may be given only in terms of density matrices, and may be obtained by summation over all possible α in the second formula from Eq. (5.6). This results in the density matrix

$$\begin{aligned} \rho_T &= \int d\alpha\, \rho_T^\alpha \\ &= \int d\alpha\, U_T^\alpha \rho_0 \,(U_T^\alpha)^\dagger \,. \end{aligned} \tag{5.8}$$

This matrix has a trace of unity (for an arbitrary initial density matrix ρ_0 having a unity trace) only if the following *generalized unitarity condition* is satisfied:

$$\int d\alpha\ (U_T^\alpha)^\dagger\ U_T^\alpha = 1. \tag{5.9}$$

This gives the normalization of the restricted path integral.

5.1.3. COMPLEX EFFECTIVE HAMILTONIAN

Now consider a more concrete (but yet very wide) case of continuous measurement, namely, the monitoring of an observable A. Such a measurement must give a value $a(t)$ for the observable at any instant t. Therefore a measurement readout may be presented by the function

$$[a] = \{a(t)|0 \le t \le T\}. \tag{5.10}$$

Because of the finite resolution of the measurement the actual value of the observable A at the instant t may differ from $a(t)$. Therefore the measurement readout, although indicated by the curve $[a]$, must actually be represented by a certain corridor around this curve.

Thus the restricted path integral (RPI) which represents monitoring of A has to be taken over paths in the corridor around $[a]$ (see Fig. 5.2). The width of the corridor Δa_T corresponds to the resolution of the measurement. The measurement readout may be represented by the function $[a]$ if

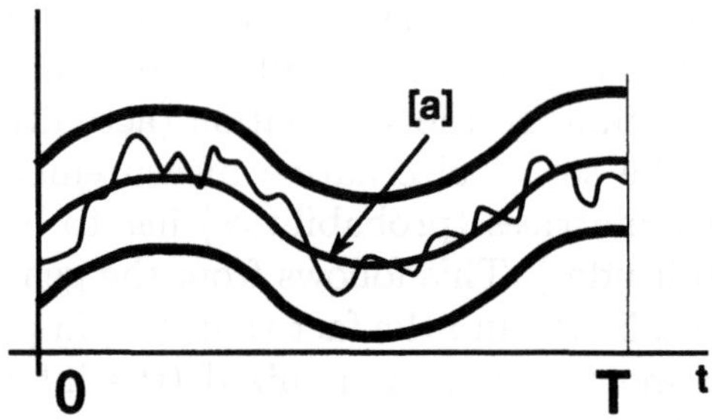

Figure 5.2. Monitoring the observable A. The path integral is restricted by a corridor around the measurement readout $[a]$.

the width is given. It is more correct, however, to consider the whole corridor of paths as an adequate image of the readout. This corridor is called a *quantum corridor.*

A more precise definition of the quantum corridor is necessary to make the RPI definite. The simplest from the mathematical point of view and, in fact, the most reasonable from the physical point of view (see Chapter 8) is the definition of the quantum corridor with the help of the Gaussian weight functional[3]

$$w_{[a]}[p,q] = \exp\left(-\kappa \int_0^T [A(p,q,t) - a(t)]^2\, dt\right). \tag{5.11}$$

The constant κ here defining the 'width' of the quantum corridor is connected with the measurement resolution:

$$\kappa = \frac{1}{T\Delta a_T^2}. \tag{5.12}$$

Notice that if κ is a constant the measurement's resolution improves with time:

$$\Delta a_T^2 \sim \frac{1}{T}.$$

This will be discussed in detail in Chapter 7. Although the Gaussian weight functional (5.11) describes the most important class of continuous measurements, we shall later consider its generalizations.

The restricted path integral with the weight (5.11) is equal to

$$\begin{aligned} U_T^{[a]}(q'',q') &= \int d[p]\, d[q]\, \exp\left\{\frac{i}{\hbar}\int_0^T (p\dot{q} - H(p,q,t))\, dt\right. \\ &\qquad \left. - \kappa \int_0^T (A(p,q,t) - a(t))^2 dt\right\} \end{aligned} \tag{5.13}$$

[3]In the general case κ may depend on t (and be included in the integrand).

It is easily seen that the RPI (5.13) with the weight functional (5.11) is equivalent to the non-restricted Feynman path integral but with an effective complex Hamiltonian (Men93; GM89; MOP91):

$$H_{[a]}(p,q,t) = H(p,q,t) - i\kappa\hbar\left(A(p,q,t) - a(t)\right)^2. \qquad (5.14)$$

It is well known that the usual Feynman path integral is equivalent to the Schrödinger equation. More precisely, this means that the propagator and the wave function evolving according to this propagator satisfy the Schrödinger equation. Now, when we have seen that the RPI is presented by the effective Hamiltonian, it is clear that the partial propagators represented by the RPI and corresponding wave functions satisfy the Schrödinger equation with the effective complex Hamiltonian (5.14):

$$\dot{\psi} = \left[-\frac{i}{\hbar}H - \kappa(A-a)^2\right]\psi. \qquad (5.15)$$

Eq. (5.15) represents the evolution of the continuously measured system under the condition that the measurement readout is given by the function $a(t)$ and that the evolution is started from the given state $\psi(0)$. The measurement's resolution is fixed by the constant κ. In the limit of small κ we have a closed system (the measurement is very small and its influence is negligible). Various measurement readouts $[a]$ arise in reality with different probability. The probability distribution is given by the probability density $p[a]$ which, according to the general formula (5.7), is equal in the given case to

$$p[a] = \langle\psi(T)|\psi(T)\rangle. \qquad (5.16)$$

It should be noticed in this connection that the probability of a single curve $[a]$ is equal to zero, so that strictly speaking the pure state evolution (5.15) is impossible. Finite probability may be ascribed only to a bundle $\mathcal{A}$ of curves $[a]$,

$$\text{Prob}([a] \in \mathcal{A}) = \int_{\mathcal{A}} p[a]\, d[a], \qquad (5.17)$$

and is expressed by the functional integral (see Sect. 5.2 and (Men88a)). The state of the system must then be represented by a density matrix, and its evolution is described by the formula analogous to Eq. (5.8):

$$\rho_T = \int_{\mathcal{A}} d[a]\, \rho_T^{[a]} = \int d[a]\, U_T^{[a]} \rho_0 \left(U_T^{[a]}\right)^{\dagger}. \qquad (5.18)$$

However, if the bundle $\mathcal{A}$ is narrow and the initial state is pure, the final state may be considered to be pure to a good approximation and represented by the effective Schrödinger equation (5.15). This is why the effective

Schrödinger equation may be efficiently used for the investigation of the behavior of continuously measured systems, just like stochastic Schrödinger equations, see Sect. 5.4.

Let us note that monitoring of an observable does not exhaust the whole scope of continuous measurements, so the concept of RPI is applicable in a wider area. Moreover, the technique of complex Hamiltonians is applicable to monitoring an observable only in the case in which a finite resolution of time is not taken into account (see Sect. 5.3).

One more generalization of the concept of monitoring is connected with the possibility that the weight functional $w_{[a]}$ may be complex, as has been argued above. Adding to the functional (5.11) a phase factor depending on $[a]$ in the simplest (linear) way, we obtain

$$w_{[a]}[p,q] = \exp\left(\int_0^T dt \left[-\kappa(t)(A-a(t))^2 - \frac{i}{\hbar}(B\,a(t)+C)\right]\right), \quad (5.19)$$

where $A = A(p,q,t)$, $B = B(p,q,t)$ and $C = C(p,q,t)$ are Hermitian observables. Just as with the functional (5.11), the complex weight functional (5.19) describes the monitoring of the observable A with a resolution determined by κ. However, with the functional (5.11) the measurement is not minimally disturbing. We shall see in Chapter 8 that this type of measurement may be realized by a long series of weak interactions. We shall see that the parameter κ in the real measurement may not only depend on time but also be an operator (with the same eigenvectors as the measured operator A). However, to a good approximation κ may be assumed to be a number.

The effective Hamiltonian has, in this case, the form

$$H_{[a]} = H + C + B\,a(t) - i\kappa\hbar\,(A-a(t))^2, \quad (5.20)$$

and the effective Schrödinger equation is following:

$$\dot{\psi} = \left[-\frac{i}{\hbar}(H+C+B\,a) - \kappa(A-a)^2\right]\psi. \quad (5.21)$$

It is evident that simultaneous continuous measurement of several observables A_n (resulting in the readouts $[a_n]$ with the resolution given by κ_n and the disturbance represented by B_n and C_n) may be described by a Hamiltonian including the sum (over n) of the corresponding additional terms in the effective Hamiltonian

$$H_{[a]} = H + \sum_n \left[C_n + B_n\,a_n(t) - i\kappa_n\hbar\,(A_n - a_n(t))^2\right], \quad (5.22)$$

or in the effective Schrödinger equation:

$$i\hbar\,\dot{\psi} = \left\{ H + \sum_n \left[C_n + B_n\,a_n(t) - i\kappa_n\hbar\,(A_n - a_n(t))^2 \right] \right\} \psi. \tag{5.23}$$

Instead, the additional terms may be included in the exponent of the weight functional (5.19).

We may choose the weight functional in an even more general form as

$$w_{[a]}[p,q] = \exp\left(-\kappa \int_0^T W(A,a)\,dt \right),$$

where $W(A,a)$ is an arbitrary complex-valued function of two multi-dimensional variables $A = \{A_n\}$, $a = \{a_n\}$, but with the square modulus $|W(A,a)|^2$ decreasing when the curves $[A]$ and $[a]$ are far from each other (the criteria of closeness could be also different). The effective Hamiltonian has then the form

$$H_{[a]} = H - i\kappa\hbar\,W(A, a(t)).$$

5.2. Derivation of the master equation

In Chapter 4 we have seen that the behavior of a particle in the process of quantum diffusion (a typical continuous measurement) is described by the master equation, the differential equation for the particle density matrix. In Sect. 5.1 a quite different description of a continuous measurement was derived. It is natural to ask how these two pictures are related. We shall see that the master equation may be derived in the framework of the RPI approach (Men94).[4] The two approaches will be shown to be essentially equivalent for non-selective description of the simple, or 'pure', type of continuous measurements, namely monitoring an observable. For simplicity we assume that the measured quantum system is one-dimensional. Generalization for an arbitrary dimension is straightforward.

5.2.1. RESTRICTED PATH INTEGRALS AND MASTER EQUATIONS

We have seen in Sect. 5.1 that the evolution of a continuously monitored system under the condition of a certain measurement output (a selective description of the measurement) is presented by the partial propagators (5.13). We can also describe the same measurement in a non-selective way, i.e., without fixing a certain measurement readout. The system is then

[4]This section is reprinted, with appropriate changes, from "Physics Letters A", Volume 196, M.Mensky, Restricted path integrals and master equations, Pages 159–167, Copyright 1994, with the permission of Elsevier Science.

described by a density matrix and evolves according to Eq. (5.8) or, in the case of monitoring,

$$\rho_T = \int d[a]\, \rho_T^{[a]} = \int d[a]\, U_T^{[a]} \rho_0 \left(U_T^{[a]}\right)^\dagger . \tag{5.24}$$

Integration over curves $[a]$ may be performed by the usual methods of path integration (Appendix A3). The following formula, typical for Gaussian path integrals, will be useful in this calculation:

$$\int d[a]\, e^{-\kappa \int_0^t dt \left((a-A'')^2 + (a-A')^2 \right)} = e^{-\frac{\kappa}{2} \int_0^t dt (A''-A')^2} . \tag{5.25}$$

With the help of this formula we can derive the following expression for a density matrix corresponding to a non-selective measurement:

$$\begin{aligned} \rho_t(q', q'') &= \int d[a]\, \langle q' |\, U_t^{[a]}\, \rho_0 \left(U_t^{[a]}\right)^\dagger | q'' \rangle \\ &= \int d[p']\, d[q']\, d[p'']\, d[q''] \\ & \quad \times \exp\left\{ -\frac{\kappa}{2} \int_0^t (A(p', q', t) - A(p'', q'', t))^2 dt \right. \\ & \quad + \frac{i}{\hbar} \int_0^t (p'\dot{q}' - H(p', q', t)) dt \\ & \quad \left. - \frac{i}{\hbar} \int_0^t (p''\dot{q}'' - H(p'', q'', t)) dt \right\} \rho_0(\bar{q}', \bar{q}'') \end{aligned} \tag{5.26}$$

(the path $[q']$ connects the points $\bar{q}'$, q', while $[q'']$ is between $\bar{q}''$ and q'').

The real (and negative) term in the exponent in this formula ensures that only those matrix elements $\langle A'' | \rho | A' \rangle$ of the density matrix emerge in its evolution for which A'' is sufficiently close to A'. This corresponds to a characteristic feature of the measurement-induced decoherence, namely, the vanishing the off-diagonal matrix elements of the density matrix.

Let us now give a brief account of the main features of another theory of quantum continuous measurements, namely, the theory based on master equations. We shall omit all details of the theory, which is rather complicated because an environment of the measured system and its interaction with the system of interest are described in this method explicitly.[5]

To derive a master equation one should consider the compound system consisting of the system of interest $\mathcal{S}$ and the measuring device, or environment, $\mathcal{E}$. Let H and H_M be Hamiltonians correspondingly of the system $\mathcal{S}$

[5]In the RPI method, instead of the detailed characteristics of the measuring setup, one need only know what information about the system this measurement supplies.

and the environment $\mathcal{E}$, while their interaction is to be described by H_I. Then the dynamics of the compound system $\mathcal{S}+\mathcal{E}$ is described in terms of its state vector $|\Psi\rangle$ or the density matrix R as follows:

$$\begin{aligned}\frac{\partial}{\partial t}|\Psi\rangle &= -\frac{i}{\hbar}(H+H_M+H_I)|\Psi\rangle,\\ \dot{R} &= -\frac{i}{\hbar}[H+H_M+H_I,R].\end{aligned}$$

The density matrix of the subsystem $\mathcal{S}$ may be obtained by tracing out degrees of freedom of the device $\mathcal{E}$:

$$\rho = \mathrm{tr}_{\mathcal{E}} R.$$

Using the equation for R and accepting some approximations, one may derive a master equation for the reduced density matrix ρ.

The master equation may have a rather complicated form reflecting details of the system and the measuring setup. However, in some regimes (corresponding to a 'pure form' of a continuous measurement of a single observable) the master equation may be of the form (WCM85; Mil87; GWM93; WM94).

$$\dot{\rho} = -\frac{i}{\hbar}[H,\rho] - \frac{\kappa}{2}[A,[A,\rho]]. \tag{5.27}$$

The first term in this equation results from the dynamical properties of the free system $\mathcal{S}$. The second term provides the decay of the off-diagonal matrix elements:

$$\frac{\partial}{\partial t}\langle A''|\rho|A'\rangle = \ldots - \frac{\kappa}{2}(A''-A')^2\langle A''|\rho|A'\rangle.$$

5.2.2. RELATION BETWEEN RPI AND MASTER EQUATIONS

We shall now prove that the density matrix (5.26) following from the RPI approach satisfies the master equation (5.27). For this aim we have to differentiate Eq. (5.26) in time using the property that the Feynman path integral satisfies Schrödinger equation (see Appendix A3). This will give

$$\dot{\rho}_t(q',q'') = \left[-\frac{i}{\hbar}(\hat{H}'-\hat{H}'') - \frac{\kappa}{2}(\widehat{A}'-\widehat{A}'')^2\right]\rho_t(q',q''), \tag{5.28}$$

where a prime means that the corresponding operator acts on the first argument, and two primes that it acts on the second argument, as follows:

$$\begin{aligned}\widehat{A}'\langle q'|\rho|q''\rangle &= A\left(-i\hbar\frac{d}{dq'},q',t\right)\langle q'|\rho|q''\rangle,\\ \widehat{A}''\langle q'|\rho|q''\rangle &= A\left(i\hbar\frac{d}{dq''},q'',t\right)\langle q'|\rho|q''\rangle.\end{aligned}$$

Eq. (5.28) is nothing else than the matrix element of the master equation (5.26) between the states $\langle q'|$ and $|q''\rangle$.

The same may be proved in another way. We can prove that the density matrix satisfying the master equation (5.27) may be represented in the form (5.26). Let us denote

$$\mathcal{H}\rho = [H,\rho], \quad \mathcal{A}\rho = [A,\rho], \quad \mathcal{B} = -\frac{i}{\hbar}\mathcal{H} - \frac{\kappa}{2}\mathcal{A}^2.$$

Then the solution to the master equation (5.27) may be written in the form

$$\begin{aligned}\rho(t) &= \mathrm{T}\exp\left[\int_0^t \mathcal{B}\,dt\right]\\ &= \mathrm{T}\exp\left[\int_0^t \left(-\frac{i}{\hbar}\mathcal{H} - \frac{\kappa}{2}\mathcal{A}^2\right)dt\right],\end{aligned}$$

where T denotes the time ordering of operators.

Let us use the representation

$$\mathrm{T}\exp\left(\int_0^t \mathcal{B}\,dt\right) = \lim_{N\to\infty} \mathrm{T}\exp(\mathcal{B}(t_N)\,\Delta t)\ldots\exp(\mathcal{B}(t_1)\,\Delta t),$$

where $t = N\Delta t$, and the r.h.s. of the equation contains N factors depending on time moments t_i dividing the interval $[0,t]$ into N equal subintervals. Let us also take into account that for small Δt one has

$$e^{(-\frac{i}{\hbar}\mathcal{H}-\frac{\kappa}{2}\mathcal{A}^2)\Delta t} = e^{-\frac{\kappa}{2}\mathcal{A}^2\Delta t}\, e^{-\frac{i}{\hbar}\mathcal{H}\Delta t}.$$

Then one can prove the path-integral representation (5.26) for the density matrix satisfying the equation (5.27).

This is a general scheme of the proof. Let us give some more details. Note that the evolution during a brief time Δt may be described as follows:

$$\begin{aligned}&\langle q'|e^{-\frac{i}{\hbar}\mathcal{H}\Delta t}\rho|q''\rangle =\\ &\int dq_1' \int dq_1'' \int \frac{dp_1'}{2\pi\hbar} \int \frac{dp_1''}{2\pi\hbar}\\ &\times e^{\frac{i}{\hbar}[p_1'(q'-q_1')-H(p_1',q_1',t)]}\,\langle q_1'|\rho|q_1''\rangle\, e^{-\frac{i}{\hbar}[p_1''(q''-q_1'')-H(p_1'',q_1'',t)]}, \qquad (5.29)\end{aligned}$$

and the exponential of a double commutator (with a brief time interval Δt) reads as

$$\langle q'|e^{-\frac{\kappa}{2}\mathcal{A}^2\Delta t}\rho|q''\rangle = e^{-\frac{\kappa}{2}(\widehat{A}'-\widehat{A}'')^2\Delta t}\langle q'|\rho|q''\rangle, \qquad (5.30)$$

where a prime in the r.h.s. means that the corresponding operator acts on the first argument, and two primes that it acts on the second argument.

Two last formulas (5.29,5.30) give us, finally:

$$\langle q'|\, e^{-\frac{i}{\hbar}\mathcal{H}\Delta t-\frac{\kappa}{2}\mathcal{A}^2)\Delta t}\, \rho\, |q''\rangle =$$
$$\int dq_1' \int dq_1'' \int \frac{dp_1'}{2\pi\hbar} \int \frac{dp_1''}{2\pi\hbar}\, e^{-\frac{\kappa}{2}\left[A(p_1',q',t)-A(p_1'',q'',t)\right]^2\Delta t}$$
$$\times e^{\frac{i}{\hbar}\left[p_1'(q'-q_1')-H(p_1',q_1',t)\right]}\, \langle q_1'|\rho|q_1''\rangle\, e^{-\frac{i}{\hbar}\left[p_1''(q''-q_1'')-H(p_1'',q_1'',t)\right]}. \quad (5.31)$$

This expression is nothing else than the first step (in time) in the path integral (5.26). Combining large number of such steps (and letting the number of steps tend to infinity) one is led to the path integral (5.26). The measure in the space of paths $[p, q]$ will automatically arise having the following form:[6]

$$d[p]d[q] = \prod_{t'=0}^{t} \frac{dp(t')}{2\pi\hbar} \prod_{t'=0}^{t} dq(t').$$

In such a way one can prove the RPI representation (5.26) for the density matrix satisfying the master equation (5.27). This completes the proof that a non-selective RPI description is equivalent to the master equation for monitoring of the observable $A(p, q)$. The generalization to higher-dimensional systems and to monitoring many observables is straightforward.

If we restrict ourselves to such continuous measurements which are performed gradually, step by step, in time (more precisely, if the measurement on a finite time interval can be resolved as a product of the measurements performed in sub intervals), then all such measurements can be represented as monitoring some observables. RPI and master equation theories are completely equivalent for non-selective description of such measurements. However, the RPI theory is more general because it is applicable to selective description as well as to continuous measurements which are 'integral in time', i.e., cannot be represented as monitoring performed step by step.

5.2.3. NON-MINIMALLY DISTURBING MONITORING

At the end of Sect. 5.1 we argued that a non-minimally disturbing continuous measurement of an observable A may be described by a complex weight functional (5.19) leading to the effective Hamiltonian (5.20) and the effective Schrödinger equation (5.21). The corresponding master equation, more complicated than Eq. (5.27), can be derived in an analogous way.

[6]The 'continuous product' of factors depending on t' should be understood as a limit $N \to \infty$ of the corresponding discrete product, i.e., the product of factors depending on discrete instants t_i separated by intervals Δt between them.

Instead of Eq. (5.25) we then need the following formula from the theory of Gaussian path integrals (see Appendix A3):

$$\int d[a]\, e^{\int_0^t dt\left[-\kappa\left((a-A'')^2+(a-A')^2\right)-\frac{i}{\hbar}a\,(B'-B'')\right]}$$

$$= e^{\int_0^t dt\left[-\frac{\kappa}{2}(A''-A')^2-\frac{1}{8\kappa\hbar}(B'-B'')^2-\frac{i}{2\hbar}(A''+A')(B'-B'')\right]}. \qquad (5.32)$$

The density matrix following from the RPI approach is (instead of Eq. (5.26))

$$\rho_t(q',q'') = \int d[p']\, d[q']\, d[p'']\, d[q'']\;\; \rho_0(\overline{q}',\overline{q}'')$$
$$\times \exp\left\{\int_0^t dt\left[\frac{i}{\hbar}(p'\dot{q}'-H') - \frac{i}{\hbar}(p''\dot{q}''-H'')\right.\right.$$
$$-\frac{\kappa}{2}(A'-A'')^2 - \frac{1}{8\kappa\hbar}(B'-B'')^2$$
$$\left.\left. - \frac{i}{2\hbar}(A'+A'')(B'-B'')\right]\right\} \qquad (5.33)$$

where we have denoted $H' = H(p',q',t)$, $H'' = H(p'',q'',t)$, and the same for the observables A, B and C.

Differentiating expression (5.33) with respect to time gives the master equation for the density matrix. In its derivation one has to notice that the product of two observables in the integrand of the path integral corresponds to the symmetrized product of the corresponding operators. For example, the product $2A(p',q',t)B(p',q',t)$ in the path integral gives the operator $\{A,B\} = AB + BA$ acting on the density matrix from the left (the product $2A(p'',q'',t)B(p'',q'',t)$ corresponds to the same operator acting from the right). Finally, we have the following master equation describing the evolution of the system subject to non-minimally disturbing continuous measurement:

$$\dot{\rho} \;=\; -\frac{i}{\hbar}[H+C,\rho] - \frac{\kappa}{2}[A,[A,\rho]] - \frac{1}{8\kappa\hbar}[B,[B,\rho]]$$
$$-\frac{i}{4\hbar}[\{A,B\},\rho] + \frac{i}{2\hbar}(A\rho B + B\rho A). \qquad (5.34)$$

5.3. Monitoring with finite time resolution

Monitoring of an observable is defined above with the help of the Gaussian corridor (5.11). In Sect. 4.3 it was shown that this definition leads to an equation which can be derived from a concrete model of continuous measurement. However, this definition is valid only in a certain approximation. It corresponds to the assumption that the measuring device (or the measuring medium) resolves time moments with absolute accuracy.

This is a good approximation in most cases, and we shall apply it later (Chapter 7) to describe the monitoring of energy in a quite realistic way. However, the restricted-path-integral approach is more general and allows for better approximation. We shall consider it here, assuming that the resolution of time is finite (see (Men97b)).[7] After discussing the phenomenological RPI (quantum corridors) approach in this case we shall consider a concrete model of measurement leading to the same result. It will be essentially the model of quantum diffusion (Sect. 4.3), but the finite relaxation time of the atoms in the medium will be taken into account.

5.3.1. QUANTUM CORRIDORS FOR A FINITE TIME RESOLUTION

It has been assumed in the RPI theory of monitoring developed in Sect. 5.1 that time is measured precisely. This means that the number $a(t)$ is an estimate, supplied by the measurement, of a value $A(t)$ of the observable A at a precisely known instant t. A real device gives an estimate of the observable A over some finite time interval of a duration, say, τ. We shall take this into account, referring to doing so as 'measuring time with the resolution τ'.

Let us first summarize the results of the preceding section, referring to the case in which time is measured precisely. Then the corresponding formulas will be generalized.

If time is measured pricisely, the non-selective description of the monitoring is given by the density matrix ρ_t defined by (5.24) and satisfying (Sect. 5.2) the equation

$$\dot{\rho} = -\frac{i}{\hbar}[H, \rho] - \frac{\kappa}{2}[A, [A, \rho]]. \tag{5.35}$$

The influence of the measuring device (measuring medium) on the measured system is described by an *influence functional* $W[p, q|p', q']$ (FV63) in the sense that the 'super-propagator' describing the evolution of the density matrix

$$\rho_T(q, q') = \int \mathrm{d}q_0 \mathrm{d}q_0' \, U(q, q'|q_0, q_0')\rho_0(q_0, q_0') \tag{5.36}$$

may be represented in the form of a double path integral:

$$\begin{aligned} U(q, q'|q_0, q_0') &= \int_q^{q_0} \mathrm{d}[p, q] \int_{q'}^{q_0'} \mathrm{d}[p', q'] \, W[p, q|p', q'] \\ &\times \exp\left[\frac{i}{\hbar}\int_0^T [(p\dot{q} - H(p, q, t)) - (p'\dot{q}' - H(p', q', t))]\right] \end{aligned} \tag{5.37}$$

[7]This section is reprinted, with appropriate changes, from "Physics Letters A", Volume 231, M.Mensky, Finite resolution of time in continuous measurements: phenomenology and the model, Pages 1–8, Copyright 1997, with the permission of Elsevier Science.

The influence functional may easily be derived from Eq. (5.24), and has the form

$$W[p, q|p', q'] = \int \mathrm{d}[a]\, w_{[a]}[p, q]\, w_{[a]}[p', q']. \tag{5.38}$$

With the weight functional (5.11) this gives, up to an inessential numerical factor,

$$W[p, q|p', q'] = \exp\left[-\frac{\kappa}{2}\int_0^T \mathrm{d}t\, (A(p, q, t) - A(p', q', t))^2\right]. \tag{5.39}$$

Now go over to the measurement in which time is measured with finite resolution. To be more concrete, let $a(t)$ be an estimate, resulting from the measurement, of the entity

$$\overline{A}(t) = \langle A\rangle_t = \int \mathrm{d}t'\, \Pi_t(t')\, A(t') \tag{5.40}$$

defined by an appropriate 'formfactor' Π_t (depending on the type of the measuring device). Then, instead of Eq. (5.11), the weight functional for restricting the path integral should be defined as follows:

$$\tilde{w}_{[a]}[p, q] = \exp\left[-\kappa\int_0^T (a(t) - \overline{A}(t))^2\, \mathrm{d}t\right]. \tag{5.41}$$

We arrive therefore, instead of at Eq. (5.13), at the following expression for the partial propagator:

$$U_T^{[a]}(q'', q') = \int_{q'}^{q''} \mathrm{d}[p, q] \exp\left\{\frac{i}{\hbar}\int_0^T (p\dot{q} - H)\mathrm{d}t - \kappa\int_0^T (\overline{A}(t) - a(t))^2 \mathrm{d}t\right\}. \tag{5.42}$$

It differs radically from Eq. (5.13) in that $\overline{A}(t)$ depends on $A(t')$ for different time moments t'. Therefore, the description of the evolution is *'not local in time'* and cannot be reduced to an effective Hamiltonian analogously to Eqs. (5.14, 5.15).

The partial propagators (5.42) describe an evolution of the measured system selectively, i.e., with the measurement output $[a]$ taken into account. A non-selective description is given by the general formula (5.8) or (5.24) resulting, in the present case, in the influence functional of the form

$$\begin{aligned}\widetilde{W}[p, q|p', q'] &= \int \mathrm{d}[a]\, \tilde{w}_{[a]}[p, q]\, \tilde{w}_{[a]}[p', q'] \\ &= \exp\left[-\frac{\kappa}{2}\int_0^T \mathrm{d}t\, (\overline{A}(p, q, t) - \overline{A}(p', q', t))^2\right].\end{aligned} \tag{5.43}$$

The measure $\mathrm{d}[a]$ of integration over measurement readouts has to be chosen in such a way that the generalized unitarity Eq. (5.9) be valid. Eq. (5.43) is valid with the conventional functional measure $\mathrm{d}[a] \sim \prod_t \mathrm{d}a(t)$ which provides the generalized unitarity either for a linear system (Men88a) or for a system with a not too strong nonlinearity and a measurement with a not too high resolution. In the general case a weight has to be included in the measure $\mathrm{d}[a]$.

The influence functional (5.43) enables one to describe the evolution of the measured system by Eqs. (5.36, 5.37) (with $\widetilde{W}$ instead of W), but no differential equation in time (analogous to Eq. (5.35)) exists for the resulting density matrix. This is a consequence of non-locality in time.

In the rest of the section we shall consider a concrete model of the monitoring of the position of a particle to verify that it actually leads to an influence functional of the form Eq. (5.43).

5.3.2. FINITE TIME RESOLUTION IN THE MODEL

The model of quantum diffusion considered in Sect. 4.2 consists of a particle in some medium. The 'atoms' of the medium are modelled as oscillators not interacting with each other. Decoherence of the particle is caused in this case by interaction of the particle with internal degrees of freedom of the atoms (presented as degrees of freedom of the oscillators).

The measuring medium in this model consists of atoms at the nodes of a cubic lattice, each atom being modelled as an oscillator. We shall represent each atom by a family of oscillators of different frequencies. However, let us begin with the case of a single oscillator with the frequency ω as in Sect. 4.2. The Hamiltonian of the interaction between the oscillator and the particle is taken in Sect. 4.2 as

$$H_{\mathrm{int}} = \sum_k H_k = \sum_k \gamma_\omega q_k \exp\left[-\frac{(\mathbf{r}-\mathbf{c}_k)^2}{l^2}\right].$$

Here $\mathbf{r}$ is a position of the measured particle while q_k is the canonical coordinate of the kth oscillator, $\mathbf{c}_k$ its location, γ_ω the interaction constant, and the length l characterizes the range of the interaction. The interaction of the particle with each of the atoms (oscillators) may be considered as a measurement of its location with the precision l.

To solve this model the discrete lattice of atoms was replaced in Sect. 4.2 by a continuous distribution of them with a constant density n. This enables one to calculate the influence functional describing the influence of the medium on the particle:

$$W[\mathbf{r}_1, \mathbf{r}_2] =$$
$$\exp\left\{ - \int_0^T \mathrm{d}t \int_0^T \mathrm{d}t' \, \nu(\omega) \cos\omega(t-t') \left[\exp\left(- \frac{(\mathbf{r}_1(t) - \mathbf{r}_1(t'))^2}{2l^2} \right) \right.\right.$$
$$\left.\left. + \exp\left(- \frac{(\mathbf{r}_2(t) - \mathbf{r}_2(t'))^2}{2l^2} \right) - 2\exp\left(- \frac{(\mathbf{r}_2(t) - \mathbf{r}_1(t'))^2}{2l^2} \right) \right] \right\}, \quad (5.44)$$

where

$$\nu(\omega) = n \left(\frac{\pi l^2}{2} \right)^{3/2} \frac{\gamma_\omega^2}{4\hbar m \omega}. \quad (5.45)$$

This simple model is not realistic enough because it does not describe measuring of time: the state of an oscillator after the interaction with the particle contains no information about the time of the interaction. To describe the measurement of time it was assumed in Sect. 4.2 that the interaction of each oscillator is turned on for a brief time. Let us now consider a more realistic model for the measurement of time.

For this goal, assume that each atom has a more complicated internal structure allowing the information about the time of the interaction to be recorded in its state. To give a model of the internal structure of the atom let us present it as a family of oscillators with different frequencies. The information about the time of interaction with the particle is now recorded in phase relations between different oscillators of the same 'atom'.

Going over to the calculation, we now have to integrate the exponent in Eq. (5.44) over the frequencies ω of the oscillators forming the model of an atom. The influence functional takes the form

$$W[\mathbf{r}_1, \mathbf{r}_2] =$$
$$\exp\left\{ - \frac{\kappa l^2}{2} \int_0^T \mathrm{d}t \int_0^T \mathrm{d}t' \, \Pi(t-t') \left[\exp\left(- \frac{(\mathbf{r}_1(t) - \mathbf{r}_1(t'))^2}{2l^2} \right) \right.\right.$$
$$\left.\left. + \exp\left(- \frac{(\mathbf{r}_2(t) - \mathbf{r}_2(t'))^2}{2l^2} \right) - 2\exp\left(- \frac{(\mathbf{r}_2(t) - \mathbf{r}_1(t'))^2}{2l^2} \right) \right] \right\} \quad (5.46)$$

with

$$\Pi(t) = \frac{2}{\kappa l^2} \int \nu(\omega) \cos\omega t \mathrm{d}\omega, \qquad \int \Pi(t) \, \mathrm{d}t = 1. \quad (5.47)$$

Let the width of the 'formfactor' $\Pi(t)$ be of the order of τ. Then for not too large an initial energy of the particle, $E_0 \ll Ml^2/2\tau^2$, the exponentials in the square brackets may be evaluated up to the first order to give

$$W[\mathbf{r}_1, \mathbf{r}_2] = \exp\left\{ -\frac{\kappa}{4} \int_0^T dt' \int_0^T dt'' \, \Pi(t'-t'') \right. \tag{5.48}$$
$$\left. \times \left[2(\mathbf{r}_2(t') - \mathbf{r}_1(t''))^2 - (\mathbf{r}_1(t') - \mathbf{r}_1(t''))^2 - (\mathbf{r}_2(t') - \mathbf{r}_2(t''))^2 \right] \right\}$$

Let the function $\Pi(t)$ be symmetrical. Then it may be expressed through the function of two arguments $\Pi_t(t')$ (depending only on the difference $|t-t'|$):

$$\Pi(t'-t'') = \int dt \, \Pi_t(t') \, \Pi_t(t''), \qquad \int \Pi_t(t') dt' = 1. \tag{5.49}$$

Substitute now this expression in Eq. (5.48) and make use of the relation

$$\begin{aligned} \int dt' \int dt'' & \\ & \times \Pi_t(t') \, \Pi_t(t'') \\ & \times \left[2(\mathbf{r}_2(t') - \mathbf{r}_1(t''))^2 - (\mathbf{r}_1(t') - \mathbf{r}_1(t''))^2 - (\mathbf{r}_2(t') - \mathbf{r}_2(t''))^2 \right] \\ = \; & 2 \int dt \, [\langle \mathbf{r}_2 \rangle_t - \langle \mathbf{r}_1 \rangle_t]^2 \end{aligned} \tag{5.50}$$

where the notation (5.40) is exploited. Then we immediately see that the influence functional has the form

$$W[\mathbf{r}, \mathbf{r}'] = \exp\left[-\frac{\kappa}{2} \int_0^T dt \, (\langle \mathbf{r} \rangle_t - \langle \mathbf{r}' \rangle_t)^2 \right] \tag{5.51}$$

in accord with Eq. (5.43). Thus the prediction of the phenomenological RPI approach coincides with what follows from the concrete model of the measurement even in the case where a finite resolution of time is taken into account.

5.3.3. CONCLUSION

We have shown, both in the RPI phenomenological approach and in the framework of the concrete model of a continuous measurement, that the finiteness of the resolution in measuring time moments leads to the replacement of the value of an observable $A(t)$ by its 'time coarse graining', $\overline{A}(t) = \langle A \rangle_t$, in the expression for the influence functional. In the special case where the position $\mathbf{r}(t)$ of a particle is continuously measured, $\mathbf{r}(t)$ has to be replaced by $\langle \mathbf{r} \rangle_t$. As a result the time evolution of the measured system cannot be represented by a differential equation in time. Neither the master equation Eq. (5.35) nor the Schrödinger equation with a complex Hamiltonian Eq. (5.15) is correct in this case.

Time coarse graining is a characteristic of the 'measuring device' or 'measuring medium'. It characterizes inertial properties of the measuring setup. It is evident that the effect of time coarse graining is negligible if all physically relevant frequencies are lower than the inverse period of coarse graining: $\Omega \ll \tau^{-1}$. In this case the master equation and the Schrödinger equation with a complex Hamiltonian are correct. Physically this means that inertial properties of the measuring medium are negligible.

From the mathematical point of view the assumption of an absolute time resolution is called Markovian approximation. The more precise approach taking into account finite time resolution considered above is, in fact, non-Markovian.

5.4. Other Approaches to Continuous Measurements

In addition to the RPI approach few other phenomenological approaches exist leading to the description of continuous quantum measurements. The most important are the following:

- An equation for the density matrix of a continuously measured system (Lindblad equation);
- The stochastic Schrödinger equation for a continuously measured system;
- The 'consistent histories' approach having a technical similarity to the RPI method but essentially different from it.

A special case of the Lindblad equation has been discussed in detail in Sects. 5.2, 5.3. In what follows we shall discuss in some detail the remaining two of them and give a brief review and comparison of all the approaches, including RPI.

5.4.1. REVIEW OF PHENOMENOLOGICAL APPROACHES

The behavior of a quantum system subject to continuous measurement can be derived by considering the model of the environment. It is possible, however, to do without such model, taking the effects of the environment into account implicitly in accordance with the scheme

System $\leftrightarrow$

We shall discuss various variants of this phenomenological description of continuous quantum measurements, paying special attention to the method of restricted path integrals.

A continuous quantum measurement can be approximated by a series of repeated instantaneous measurements, each of which is described by a set of von Neumann projectors. Evolution of the system between two measurements is described by the Schrödinger equation, or (which is equivalent) by the unitary evolution operator. This gives us the following law of evolution between the times t_0 and $t = t_N$:

$$\begin{aligned} |\psi(t)\rangle &= U(t_N, t_{N-1}) P_{i_{N-1}} U(t_{N-1}, t_{N-2}) \dots \\ &\quad \dots U(t_3, t_2) P_{i_2} U(t_2, t_1) P_{i_1} U(t_1, t_0) |\psi(t_0)\rangle \end{aligned} \tag{5.52}$$

The result of a series of measurements is represented by a sequence of numbers $\{i_1, i_2, \dots, i_{N-1}\}$. We see that the evolution of the measured system depends on these numbers, that is, on the result of the series of measurements. The description can be made more realistic if we represent the instantaneous measurement not by von Neumann projectors, but rather by their generalizations, positive operators (see Sect. 2.3). We then have a sequence of instantaneous fuzzy measurements separated by intervals of free evolution:

$$\begin{aligned} |\psi(t)\rangle &= U(t_N, t_{N-1}) R_{i_{N-1}} U(t_{N-1}, t_{N-2}) \dots \\ &\quad \dots U(t_3, t_2) R_{i_2} U(t_2, t_1) R_{i_1} U(t_1, t_0) |\psi(t_0)\rangle \end{aligned} \tag{5.53}$$

The spectrum of measurement outputs in each instantaneous measurement can be continuous. The simplest situation, when intervals of free evolution may be neglected, has been discussed in Sect. 2.4.

Passing to the limit of very frequently repeated measurements one can obtain a tool for studying continuous quantum measurements; see, for example, (BLP82; Men83; Kha88). The resulting approach is phenomenological and does not require an explicit model of the measuring medium. The continuously measured system is then assumed to be open, and the effects of the environment are taken into account in an implicit way.

There are quite a few phenomenological approaches which lead straight to the continuous measurement, without first representing it as a repeated measurement with subsequent passage to the limit. The RPI approach discussed in detail in Sect. 5.1 reduces the description of a continuously measured system $\mathcal{S}$ to Schrödinger equation with complex Hamiltonian

$$|\dot{\psi}\rangle = \left[-\frac{i}{\hbar} H - \kappa (A - a(t))^2 \right] |\psi\rangle. \tag{5.54}$$

By $a(t)$ we here denote the value of the A at the time t found in the course of its continuous measurement. It is important that this approach is selective. Because of the explicit account of the measurement readout $[a]$ it describes an individual system, not a quantum ensemble of systems.

The principal ones of the other phenomenological approaches are following.

– An equation for the density matrix for a continuously measured (open) system (the master equation). In the simplest case it reads

$$\dot{\rho} = -\frac{i}{\hbar}[H, \rho] - \frac{1}{2}\kappa[A, [A, \rho]], \tag{5.55}$$

where H is the Hamiltonian of the measured system, A is the continuously measured observable of the system, that is, the observable the information whereof is 'recorded' in the state of the environment, and the constant κ characterizes the strength of interaction between the measured system and its environment. In Sect. 4.2 we saw that the equation of this form can be derived from the model of interaction between the system and its environment. In Sect. 5.2 it has been derived from the RPI approach in the case when the continuous monitoring of the observable A is minimally disturbing, i.e., represented by a real weight functional (see Sect. 5.1). For a non-minimally disturbing monitoring (a complex functional) the RPI approach gives a more complicated master equation (5.33). Equation (5.55) is a special case of the equation

$$\dot{\rho} = -\frac{i}{\hbar}[H, \rho] + \sum_{\nu}([L_\nu\rho, L_\nu^\dagger] + [L_\nu, \rho L_\nu^\dagger]). \tag{5.56}$$

In (Lin76) this equation was derived by Lindblad for an open system without using any particular model, under the assumption of the Markovian nature of the process. Interpretation of the special case (5.55) of the Lindblad equation as describing a minimally disturbing continuous measurement can be corroborated either by the models of measurement, or with the aid of restricted path integrals (see next paragraph).

– The stochastic Schrödinger equation for a continuously measured system

$$d|\psi\rangle = \left[-\frac{i}{\hbar}H - \kappa(A - \langle A\rangle)^2\right]|\psi\rangle dt + \sqrt{2\kappa}(A - \langle A\rangle)|\psi\rangle dw \tag{5.57}$$

can be derived (see Sect. 5.4.2) from the equation with complex Hamiltonian (5.54). In the stochastic equation w is a random function of the type of Brownian walk (white noise) which characterizes the effects of the environment. The differential of this variable which enters this equation satisfies the condition $dw^2 = dt$, which reflects the properties of Brownian motion (the mean deviation is proportional to the square root of the time interval). Other stochastic equations (Gis89; Bel89; DGHI95) have also been proposed from which the master equation follows. The stochastic equation cannot be unambiguously derived from the Lindblad equation. The advantage of Eq. (5.57) is that it follows from the equation with complex Hamiltonian (5.54), which in turn can be derived from first principles. In (GRW86) and subsequent publications of the same authors the noise in the stochastic equation was interpreted in a different fashion — not as an effect of the environment, but rather as an independent fundamental physical process, called spontaneous localization.

The relationship between the various phenomenological approaches to continuous quantum measurements can be illustrated by the following table:

Quantum mechanics of open measured systems

Selective description		Non-selective description
⇓		⇓
Feynman Quantum Mechanics: Feynman 1948	↘	
↓		
Restricted Path Integrals: Mensky 1979	→	**Influence functional:** Feynman & Vernon 1963
↓		
Complex Hamiltonians: Golubtsova & Mensky 1989; Mensky, Onofrio & Presilla 1991	↘	↓
↓		
Stochastic Equations Diosi 1989; Gisin 1989; Belavkin 1989	→	**Master equation:** Lindblad 1976

In this table we emphasize the distinction between the selective and non-selective descriptions of continuous quantum measurements. The selective description is more detailed. It portrays the evolution of the measured system for only one out of many alternative results of measurement. In this description the state of the measured system remains pure (if, of course, it had been pure prior to the measurement). In the equation with a complex Hamiltonian the alternative is given by the function $a(t)$, which has a straightforward physical meaning, namely, it is the result of monitoring the observable A. In the stochastic equation the alternative is given by the random function w.

The non-selective description represents evolution of the measured system irrespective of the measurement's readout. This description takes into account all possible readouts, and the actual readout is assumed to be not known. Accordingly, the non-selective description (the equation for the density matrix, or master equation) can be derived from the selective description by carrying out summation with respect to the alternatives (see Sect. 5.2). The transition back from non-selective (less detailed) to selective (more detailed) description is ambiguous and requires additional assumptions for removing the uncertainty. The logical links between the approaches, the possibilities for deriving one approach from another, are shown on the diagram by arrows (see also Sect. 5.4.2).

All the above are directly concerned with fuzzy continuous measurements (although the fuzziness may be minor). There are techniques applicable instead to series of projecting measurements. Thus Carmichael worked out (Car93) an interesting mathematical technique resembling that of the path integral but applicable to a series of projecting instantaneous measurements, such as counting of photons. He called it the technique of 'quantum trajectories'.

5.4.2. STOCHASTIC EQUATION

The Schrödinger equation with complex Hamiltonian

$$|\dot{\psi}\rangle = -\frac{i}{\hbar}H_{[a]}|\psi\rangle = \left[-\frac{i}{\hbar}H - \kappa(A - a(t))^2\right]|\psi\rangle, \tag{5.58}$$

may be rewritten in the form of a stochastic Schrödinger equation, as is shown in (POT96; AKS97). To this end introduce new variables instead of the function $a(t)$ and the state vector $\psi(t)$, in the following way:

$$a = c + \frac{\xi}{\sqrt{2\kappa}}, \quad \Psi(t) = \exp\left(\frac{1}{2}\int_0^t dt\,\xi^2\right)\psi(0), \tag{5.59}$$

where $c(t)$ will be chosen later. Then the equation (5.58) takes the form

$$|\dot{\Psi}\rangle = \left[-\frac{i}{\hbar}H - \kappa(A - c)^2 + \sqrt{2\kappa}(A - c)\xi\right]|\Psi\rangle, \tag{5.60}$$

or, in terms of the new variable w defined as $dw = \xi dt$,

$$d|\Psi\rangle = \left[-\frac{i}{\hbar}H - \kappa(A - c)^2\right]|\Psi\rangle dt + \sqrt{2\kappa}(A - c)|\Psi\rangle dw. \tag{5.61}$$

For the norm of the vector $|\Psi\rangle$ to be conserved,

$$(\langle\Psi| + \langle d\Psi|)(|\Psi\rangle + |d\Psi\rangle) = \langle\Psi|\Psi\rangle, \tag{5.62}$$

it is enough to take

$$dw^2 = dt, \quad c = \langle\Psi|A|\Psi\rangle. \tag{5.63}$$

The equation (5.61) with the condition (5.63) is called the stochastic Schrödinger equation. It is non-linear, contrary to the linear equation (5.58). The letter w denotes here a random function (noise) representing the influence of the measuring medium on the measured system. The statistics of the noise is determined by the probability distribution

$$p[w] = \langle\psi|\psi\rangle = \exp\left(-\frac{1}{2}\int_0^t \dot{w}^2(t)dt\right). \tag{5.64}$$

This is a so called *white noise* or Brownian motion. The change of w during a small time interval Δt is then given by the distribution

$$p(\Delta w) = \frac{1}{\sqrt{2\pi\Delta t}}\exp\left(-\frac{1}{2}\frac{\Delta w^2}{\Delta t}\right), \tag{5.65}$$

hence the mean square of the deflection is proportional to time:

$$\overline{\Delta w^2} = \int p(\Delta w)\Delta w^2 d\Delta w = \Delta t. \tag{5.66}$$

The last formula gives a physical interpretation of the unusual relation $dw^2 = dt$, accepted in the stochastic calculus in the framework of the so called *Ito formalism* (Ito51).

The stochastic Schrödinger equation (5.61) is thus derived from the theory of quantum continuous measurements based on RPI or the Schrödinger equation with complex Hamiltonian. Some other stochastic equations were also proposed for the description of continuous quantum measurements (Gis89; Dio89; Bel89). The necessary condition in each case is that the density matrix describing the same measurement in the non-selective way satisfy the master equation. This occurs for Eq. (5.61) because Eq. (5.55) is a consequence of the RPI approach (see Sect. 5.2).

5.4.3. CONSISTENT HISTORIES

Some years ago a direction in quantum mechanics based on the concept of *consistent histories* arose (Gri84; Omn90; GMH90; GMH93). The aim was to try and answer the question of how a quantum system acquires classical features. In other words, the aim was the analysis of the phenomenon of decoherence.

The consistent histories approach, though essentially different from the method of restricted path integrals (RPI), has some technical similarity with the latter. The difference is that consistent histories are used for the description of a closed quantum system (with the aim of revealing its classical features) while the RPI method deals with an open system, taking its environment into account in an implicit way. Let us quite briefly describe the consistent histories approach following the work of Gell-Mann and Hartle (GMH90). We shall on purpose make use of the same notations as in the RPI approach to underline the technical similarity, although the corresponding concepts, as we shall see, will be quite different.

A history $\alpha = \{i_1, i_2, \ldots, i_{N-1}\}$ is defined in (GMH90) as a chain of projectors specifying the states of the system in successive time moments. These projectors determine the evolution operator

$$\begin{aligned} U_\alpha &= U(t_N, t_{N-1}) P_{i_{N-1}} U(t_{N-1}, t_{N-2}) \ldots \\ &\quad \ldots U(t_3, t_2) P_{i_2} U(t_2, t_1) P_{i_1} U(t_1, t_0) \end{aligned} \qquad (5.67)$$

analogous to one used in the formula (5.52) for description of a series of instantaneous measurements. However, in the consistent histories approach projectors serve not for the description of a measurement but only for the analysis of the free evolution of the system without any external influence. The operator U_α is only one of many quantum alternatives contributing the evolution. The total evolution operator is equal to the sum

$$U = \sum_\alpha U_\alpha \qquad (5.68)$$

over all possible histories i.e.,, over all possible choices of the projectors for each of the time moments.[8]

Histories in the sense of Gell-Mann and Hartle are direct analogues of Feynman paths which determine the evolution of a closed system only when they are taken as a whole (in the path integral). Contrary to the case of individual paths, histories are more rough (more coarse-grained) alternatives, each one of them including many paths.

With each history Gell-Mann and Hartle associate a 'probability'

$$P_\alpha = \mathrm{tr}(U_\alpha \rho U_\alpha^\dagger). \qquad (5.69)$$

Gell-Mann and Hartle underline that interpretation of this entity as a probability is not necessarily correct, and their further analysis aims to answer the question when this interpretation is justified.

[8]If the operator (5.67) had represented a measurement, as is assumed in Eq. (5.52), then it could be associated with a classical alternative, so no sense would be in summing over such operators.

Together with the given set of histories $\{\alpha\}$ the authors also consider the sets of histories obtained from the given one by the procedure of *coarse graining* in which the transition is performed to wider projectors or/and to more rarely projecting (not at each of the chosen time moments). Each history in the rougher set may be presented as a sum of histories from the finer set (a rough history consists of the fine histories as of alternatives). The corresponding evolution operators are also connected by the operation of summing up:

$$\beta = \sum_{\alpha\in\beta} \alpha, \quad U_\beta = \sum_{\alpha\in\beta} U_\alpha. \tag{5.70}$$

However, the 'probabilities' corresponding to rougher histories

$$P_\beta = \mathrm{tr}(U_\beta \rho U_\beta^\dagger) \tag{5.71}$$

are not obliged to be sums of 'probabilities' of fine histories.

The following consideration is the key one for the whole approach since it leads to the 'condition of consistency' as a *necessary criterion* that the description of the system in terms of alternatives $\{\alpha\}$ is classical.

A sufficiently rough description of a quantum system loses the specific quantum features and becomes purely classical. The alternatives $\{\alpha\}$ give a coarse grained description of the system, and if the degree of coarse graining is enough then the alternatives may be considered as classical ones. In this case the entities (5.69) must be considered as probabilities. The conventional rule of summing probabilities may be applied to them in this case.

If alternatives from the set $\{\alpha\}$ may be considered classical then any even rougher set of alternatives is also classical. Therefore the $\{\beta\}$ is also classical, and the following *probabilities summing rule* must be valid:

$$P_\beta = \sum_{\alpha\in\beta} P_\alpha. \tag{5.72}$$

Thus, if the alternatives $\{\alpha\}$ are classical then in the transition to the rougher alternatives $\{\beta\}$ not only the amplitude summing rule (5.70) but also the probabilities summing rule (5.72) must be valid. A simple mathematical analysis shows that the requirement (5.72) is ensured if the *decoherence functional*

$$C_{\alpha\alpha'} = \mathrm{tr}(U_\alpha \rho U_{\alpha'}^\dagger) \tag{5.73}$$

satisfies the *consistency condition*

$$C_{\alpha\alpha'} + C_{\alpha\alpha'}^* = 0, \quad \text{for} \quad \alpha \neq \alpha' \tag{5.74}$$

Therefore the consistency condition (5.74) is necessary for the set of histories $\{\alpha\}$ to be interpreted as a set of classical alternatives.

It has been shown that the consistency condition is approximately valid in the 'situation of measurement', i.e., in the case where the system under consideration contains the two parts (subsystems) interacting in such way as a measuring device and a measured system must interact (see Chapter 2). However, the consistency condition turned out (PZ93) to be insufficient to provide *predictability* of the quantum system in the given degree of coarse graining, while such a property is necessary for a classical system. Therefore the consistent histories approach has not solved completely the problem of the transition from quantum to classical, though it has made the question more clear, particularly underlining the importance of the concept of coarse graining.

Returning once more to the comparison of the consistent histories and RPI approaches we have to notice that in the latter no consistency condition arises. The reason is that the system is considered in the RPI approach as an open one interacting with the given environment (although this environment is not modelled explicitly). This is why only one set of histories (or quantum corridors) $\{\alpha\}$ is used in the RPI approach, namely one presenting the influence of the given environment. No necessity arises of considering simultaneously more rough sets of histories and agree with each other the descriptions having different degrees of coarse graining. The classical character of the behavior of the measured system (its decoherence) is provided in this case not by its internal properties, but by the influence of the environment.

5.5. Comments

The idea of the RPI approach to continuously measured systems was put forward by Richard Feynman in a brief remark in his famous paper of 1948 (Fey48). Some unsuccessful attempts at developing the idea then followed (BB74; Fri76). The mistake made in these works was that the corridor which had to restrict the path integral was defined by null boundary conditions or by a repulsive (real) potential. In (Men79a; Men79b) a correct realization of Feynman's idea was suggested which made use of an imaginary 'absorbing' potential. In a long series of papers and books (see, for example, the books and reviews (Men83; Men93; Men96b; Men97a; Men98a; Men98b)) the theory of restricted path integrals has been elaborated and applied to many various measurement situations (not only in non-relativistic quantum mechanics, but also in quantum field theory). Feynman's idea was essentially extended in that not only the probability distribution of the measurement readout but also the evolution of the measured system is presented by RPI.

The theory was contributed to by other authors also (Kha81; BLP82; Cav86; Dio92). In (AV80) the original idea of Feynman was realized with the somewhat extreme and less realistic presupposition that a solitary Feynman path has to be singled out by the measurement (which resulted in a sort of Zeno paradox).

The thought of applying a complex Hamiltonian to represent the effect of restricting the path integral was put forward in (GM89), then applied in (MOP91; MOP93; OPT93) for numerical calculations of partial propagators.

A direct and physically transparent analysis of continuous measurements with the help of explicit models of continuous measurements goes back to works of Joos and Zeh (JZ85) for quantum-mechanical systems and of Milburn with collaborators (WCM85; Mil87; GWM93; WM94) for systems in quantum optics. Evolution of the system under measurement was described in these models by master equations. Independently, master equations for open systems were derived by Lindblad without models but with an assumption about the Markovian character of the process (Lin76). Recently Michael Marinov, my friend from student days in Moscow University, reminded me of his paper of 1972 (Mar72; Mar74) in which the evolution of a pure state into a mixed state and a master equation has been discussed. This work was not properly understood and valued as much as its later version (Mar85).

Various stochastic equations were proposed for the description of continuous quantum measurements in (Gis89; Dio89; Bel89). The equivalence of the complex-Hamiltonian and stochastic Schrödinger equation was proved in (POT96; AKS97).

Chapter 6

Restricted Path Integrals: General Features

Definition of the restricted-path-integral (RPI) approach has been given in Chapter 5 and its applications will be considered later on in Chapters 7, 8. Here we are going to discuss some general features of the approach which make it especially interesting from the theoretical and conceptual points of view (see also Chapter 10 for more philosophical aspects of this).

After brief general remarks in Sect. 6.1 we show in Sect. 6.2 that difficulties in the mathematical justification of Feynman path integrals are naturally overcome in a more realistic case of restricted path integrals. Then we discuss in much detail the so called 'action uncertainty principle' (Sect. 6.3) (playing role of the well known Heisenberg uncertainty principle but for continuous measurements).

Some important general features of the RPI approach were discussed in the previous book (Men93), among them a peculiar group-theoretical structure which can be found in continuous measurements. It resembles the Galilean group but with shifts of paths instead of three-dimensional translations, and with multiplication of corridors (defined at adjacent time periods) instead of time translation.

This chapter may be skipped by the reader interested only in the results closer to practical applications.

6.1. Peculiarities of the RPI (quantum corridor) approach

The RPI approach to continuous quantum measurements has some specific features which make it advantageous from the conceptual point of view:

- No postulates but the principles of quantum mechanics
- Selective description
- Dynamical role of information

As has been seen in Sect. 5.1, the theory of continuous measurements in the RPI form may be derived from the general principles of quantum mechanics without any additional postulates. The only condition is that quantum mechanics is formulated in terms of Feynman path integrals. The underlying ideology of quantum mechanics is the theory of amplitudes, which is quite important for the derivation.

It is interesting that this logic leads to continuous quantum measurements without any mention of instantaneous measurements. This is, in fact, natural because what is called 'instantaneous measurement' is not actually instantaneous. In reality such a measurement is only brief, so its duration may be neglected. Therefore continuous measurement is, in fact, primary, and an instantaneous measurement may be obtained as its limit (Cal95).

Therefore the whole quantum theory of measurements, if formulated in this way, becomes a part of quantum mechanics. *Quantum mechanics* (in the Feynman form) becomes *a closed theory*. No quantum theory of measurements as an additional system of postulates is necessary.

Another specific feature of the RPI approach is that it allows for the presentation of the measured system by its state vector (wave function), not thes density matrix. The reason is that the measurement is described selectively, i.e., the concrete readout of the measurement is taken into account. Of course, the less detailed non-selective description may also be readily derived.

A continuously measured system is open, influenced from outside. Nevertheless, it is represented in the RPI approach without explicit reference to the environment. Back influence of the environment is taken into account implicitly. All that is necessary for this in the RPI approach is specification of the information about the system which is recorded in the environment. This information is represented in the RPI approach by the quantum corridor.

Therefore the RPI (quantum corridor) approach is an *information approach to quantum measurements*. It reveals the *dynamical role of information*, the role having no analogue in classical mechanics:

– Information about a quantum system inevitably modifies its dynamics.
– The modification is completely determined by the information.

The first of these issues is well known. The second is not novel either, but it is represented in the RPI approach in the most complete and clear form. It is not novel because it is characteristic of instantaneous measurements too and is expressed by von Neumann's projection postulate: the measurement record determines, according to this postulate, the final state of the system.

From this point of view restricted path integrals may be considered to be an *analogue of the von Neumann postulate but for processes* instead of states. The evolution process (the set of Feynman paths) is projected by the RPI (by a quantum corridor) in such a way as to be in accord with the measurement readout.

The theory of restricted path integrals leads, in fact, to a new form of quantum mechanics, the *theory of open continuously measured quan-*

tum systems. In this theory the quantum system may (in the special case of monitoring) be described by a Schrödinger equation with a complex Hamiltonian. The evolution of the system in this theory has both quantum and classical features. Namely, the system evolves quantum mechanically in the quantum corridor, but the corridors obey classical statistics. This may be thought of as a universal *theory of decoherence*, the dynamics of decoherence.

The Schrödinger equation with a complex Hamiltonian or equivalent stochastic equations may be a convenient technical instrument for dealing with the dynamics of continuous measurements, or the dynamics of decoherence. However, the deep conceptual structure of quantum measurements (decoherence) is naturally formulated in the language of restricted path integrals.

6.2. RPI and foundations of path integrals

The formalism of path integrals,

$$U(q'', q') = \int d[q]\, e^{\frac{i}{\hbar}S[q]}, \tag{6.1}$$

was introduced by Richard Feynman (Fey48; FH65) as a mathematical instrument of quantum mechanics possessing great heuristic abilities. However, from the mathematical point of view this was done not completely rigorously, 'on physical level' (see Appendix A3). Later works of many authors showed that it is not so simple to give a full mathematical justification of Feynman path integrals. The difficulties arise because of the purely oscillating character of the exponential in the integrand of Eq. (6.1), i.e., because of the imaginary unit in the exponent and the action functional $S[q]$ being real.

A mathematically rigorous theory of path integrals can be found in (Kac58), but it deals mostly with the Wiener path integrals arising in statistical mechanics:

$$\int d[q]\, e^{-\frac{1}{\hbar}S[q]}, \tag{6.2}$$

These integrals are characterized by a decreasing exponential. As a result the convergence of the limiting processes used in their definition is much better. There are many papers devoted to the mathematically strict foundation of Feynman path integrals arising in quantum mechanics; see, for example, (MD72; Kla87) and references therein. The problem turns out not to be easy, and it solution rather artificial.

We shall argue (Men92c) that the root of these difficulties is an over-idealization. The system described by the propagator (6.1) is assumed closed, i.e., completely isolated. This is, in fact, impossible. If we accept a more realistic assumption of an open continuously measured system, then, according to Sect. 5.1, we have to restrict the path integral or, equivalently, to introduce an imaginary term in the action functional. Therefore the convergence of the path integral will be better and its definition better justified.

For convenience we have hitherto spoken of continuous measurement. However, it should not be thought that this necessarily means the presence of an instrument specially constructed to extract information about our system. For the treatment described in Chapter 5 to be valid it is, in fact, sufficient if there exists, in principle, the possibility of obtaining information of a given type about the behavior of the system. If such a possibility exists in principle, this means that in the process of the evolution of the system it interacts with its environment in a specific manner, leaving in the environment traces from which information about the behavior of the system can be recovered. In accordance with the laws of quantum mechanics such an interaction is necessarily associated with a back reaction of the environment on the system. The formalism of restricted path integrals (RPI) takes into account this back reaction.

It is easy to see that some information about the behavior of the system always exists. For example, any experiment with an elementary particle is made in some spatial region (in a laboratory), and it is known that the particle does not leave this region during the time of the experiment. Therefore to calculate the transition amplitude (the propagator) of the particle, it is certainly not necessary to integrate over paths which pass outside this region. One can say that there is always monitoring of the particle's position, admittedly with a very poor, but finite, resolution. It is clear that in any real situation there are analogous restrictions not only on the position of the system in the configuration space but also on other aspects of its behavior, for example, the velocity.

Of course, in the majority of such situations the restrictions are insignificant, since they lead to negligibly small corrections to the law of evolution of the system. However, for the problem of a mathematically correct definition of Feynman integrals, all, even arbitrarily small but finite, measurement effects are important.

The continuous measurement of velocity resulting in the curve

$$[v] = \{v(t)|0 \leq t \leq T\}$$

is described, according to Sect. 5.1, by the restricted path integral which can be represented by an imaginary additional term in the action or, equivalently, by a negative additional term in the exponent:

$$U_{[v]}(q'', q') = \int d[q]\, e^{\frac{i}{\hbar}S[q] - \frac{1}{\Delta v^2 T}\int_0^T (\dot{q}(t)-v(t))^2\, dt}. \tag{6.3}$$

In the majority of real situations no measurement of the velocity occurs. However, this does not mean that in these cases no information about the velocity remains. Indeed, in practice there is always information, which amounts to a restriction on the velocity. The point is that if the velocity (the energy) exceeds a certain threshold, phenomena of a qualitatively different nature (for example, disintegration of the system) must occur. If these phenomena were not observed it means that the velocity did not exceed the threshold. Denoting this threshold by V, we have for the corresponding restricted path integral:

$$U_0(q'', q') = \int d[q]\, e^{\frac{i}{\hbar}S[q]}\, e^{-\frac{1}{V^2 T}\int_0^T [\dot{q}(t)]^2\, dt}. \tag{6.4}$$

An additional cutting factor (exponential with a negative argument) in this path integral improves its convergence. In fact, the integral (6.4) may be considered as a Wiener path integral with the Wiener measure defined by the real exponential. The original exponential with an imaginary argument may be considered as a functional integrated with this Wiener measure. Contrary to the Feynman integral, the Wiener measure is defined mathematically correctly. As a result, well defined is also the Feynman integral regularized by the real exponential.

The same is valid also for a more general Eq. (6.3). In this case the path integral depends on the function $v(t)$. This is natural because the function $v(t)$ determines a classical alternative corresponding to one of the readouts of a continuous measurement. The quantum evolution represented by the (restricted) path integral depends of course on the measurement result.

The conclusion now suggests itself very obviously. Namely, the difficulties with the correct definition of the Feynman integral disappear if one gives up the idealization associated with the concept of an isolated system and takes into account the influence of an environment having the nature of a measurement. In the presence of 'measurement' of the velocity or, in a special case, 'restriction' of the velocity in the sense described above, there appears an imaginary part of the action equivalent to the introduction into the Lagrangian of a term of the form of $i\epsilon[\dot{q}(t) - v(t)]^2$ with a finite ϵ. This gives a regularized and well defined path integral.

6.3. Action uncertainty principle

The Action Uncertainty Principle (AUP) is an analogue of the usual uncertainty principle but for quantum processes instead of states. AUP enables one to specify the most probable readouts of continuous quantum measurements. Here we follow the paper (Men96a).[1]

In what follows the continuous monitoring of an observable $A(p,q,t)$ with resolution $\Delta a(t)$ is considered. The influence of the measurement process on the measured system (quantum measurement noise) is represented by an additional term $\delta F(t)A(p,q,t)$ in the Hamiltonian, where the function δF (a generalized fictitious force) is restricted by the AUP $\int |\delta F(t)|\, \Delta a(t)\, dt \lesssim \hbar$ and otherwise arbitrary. Quantum non-demolition (QND) measurements are analyzed with the help of AUP. A simple uncertainty relation for continuous quantum measurements is derived. It states that the area of a certain band in the phase space should be of the order of $\hbar$. The width of the band depends on the measurement resolution, whilst its length is determined by the deviation of the system, owed to the measurement, from its classical behavior.

6.3.1. FORMULATION OF THE PROBLEM

The evolution of a continuously measured system is described by a restricted path integral (RPI). In Sect. 5.1 we defined the RPI as a weighted path integral. Here, however, we shall prefer a simpler but more transparent form of the path integral over some restricted set of paths. Thus for the RPI in the coordinate representation we have

$$U_\alpha(q'',q') = \int_{I_\alpha} d[q]\, e^{\frac{i}{\hbar}S[q]}, \tag{6.5}$$

where I_α is the set of paths (quantum corridor) representing the measurement readout I_α. This form of RPI is applicable, for example, if we consider the monitoring of the coordinate.

In many cases the measurement readout cannot be expressed in terms of a corridor in the coordinate space. The phase space representation of the path integrals then has to be applied. The RPI then has the form

$$U_\alpha(q'',q') = \int_{J_\alpha} d[p]d[q]\, e^{\frac{i}{\hbar}\int_0^T (p\dot{q}-H(p,q,t))}, \tag{6.6}$$

where J_α is a set of paths (a quantum corridor) in the phase space corresponding to the measurement readout α.

It is more realistic to describe continuous measurements by weighted, rather than sharply restricted, path integrals. However, we shall use here the form (6.6), which is enough for calculations up to the order of magnitude, but makes the consideration easier.

[1]This section is reprinted, with appropriate changes, from "Physics Letters A", Volume 219, M.Mensky, The Action Uncertainty Principle for continuous measurements, Pages 137–144, Copyright 1996, with the permission of Elsevier Science.

The set of partial propagators (partial evolution operators) U_α describes the evolution of the measured system during the period $[0, T]$ of the measurement by Eqs. (5.6, 5.8). This law of evolution conserves the normalization ($\mathrm{tr}\rho' = 1$) provided that the set of the partial evolution operators U_α satisfies the generalized unitarity condition $\int d\alpha\, U_\alpha{}^\dagger U_\alpha = 1$. Then $\mathrm{tr}\rho_\alpha$ is a probability density of the measurement readout α.

Consider a simple situation where the position of the system in the configuration space is given ('instantaneously measured') before and after the continuous measurement. Let these positions be given by q' and q'' respectively. Then the (relative) probability density of the measurement readout α may be estimated by the square modulus $|U_\alpha|^2 = |U_\alpha(q'', q')|^2$ of the partial propagator (we shall omit hereafter the letters q', q'').

For a rough estimate of which measurement readouts α have high probability and which of them have low probability one should find the readout α corresponding to the maximum $|U_\alpha|$ (it proves to coincide with the classically predicted readout α_{class}), the set of readouts corresponding to the values $|U_\alpha|$ close to maximum, and the set of readouts for which $|U_\alpha|$ is much less than the maximum value. As will be shown below, the set of probable readouts found in this way may be characterized by a simple inequality, the Action Uncertainty Principle (AUP).

6.3.2. DERIVATION OF THE AUP

The Action Uncertainty Principle (AUP) is an inequality which allows one to specify in a simple way the subset of the measurement readouts α having comparatively high probabilities.

Monitoring the coordinate We shall begin with a simple special case: monitoring the coordinate q of a one-dimensional system. In this case the coordinate representation of the path integral may be applied. Let the measurement readout be expressed by the function $[a] = \{a(t) | 0 \le t \le T\}$ whilst the measurement's error (resolution) is characterized by $[\Delta a] = \{\Delta a(t) | 0 \le t \le T\}$. This means that α in the preceding formulas must be replaced by $[a]$, the partial propagator may be denoted by $U_{[a]}$ and the integration in Eq. (6.5) must be restricted to the paths $[q]$ lying in the corridor between $[a - \Delta a]$ and $[a + \Delta a]$.

Starting from Eq. (6.5), going over to integrating over $[\Delta q] = [q - a]$ and expanding the action in powers of $[\Delta q]$, we have

$$
\begin{aligned}
e^{-\frac{i}{\hbar}S[a]}\,U_{[a]} &= \int_{I_{[a]}} d[q]\, e^{\frac{i}{\hbar}(S[q]-S[a])} \\
&= \int_{-[\Delta a]}^{[\Delta a]} d[\Delta q]\, \exp\Big\{\frac{i}{\hbar}\Big[\int_0^T \frac{\delta S[q]}{\delta q(t)}\Delta q(t)dt \\
&\quad + \int_0^T\int_0^T \frac{\delta^2 S[q]}{\delta q(t)\delta q(t')}\Delta q(t)\Delta q(t')dtdt' + \ldots\Big]\Big\}. \qquad (6.7)
\end{aligned}
$$

The functional derivatives here are taken at $[q] = [a]$. We shall assume that the higher terms of the expansion are not important for estimating $|U_{[a]}|$ up to the order of magnitude.

To specify the most probable measurement readouts we should find out in which cases an absolute value $|U_{[a]}|$ of the amplitude $U_{[a]}$ is close to maximum. The maximum is achieved for $[a] = [a_{\text{class}}]$ (the classical trajectory) when the first functional derivative of the action $S[q]$ vanishes and the linear term in the expansion of the action is absent. For $[a]$ differing from $[a_{\text{class}}]$ the linear term is non-zero and leads to a decrease of $|U_{[a]}|$ owing to destructive interference.

We shall assume that the absolute value of the integral (6.7) decreases mainly because of the linear term in the expansion of the action. The probability of the measurement readout is then high (close to the maximum possible probability) until the linear term becomes equal to or greater than $\hbar$. In other words, the condition for the measurement readout $[a]$ to be highly probable is

$$
\max_{[\Delta q]}\left|\int_0^T \left.\frac{\delta S[q]}{\delta q(t)}\right|_{[q]=[a]} \Delta q(t)\, dt\right| \lesssim \hbar. \qquad (6.8)
$$

Here the maximum should be taken for all $[\Delta q]$ not bringing $[q]$ out of the corridor $I_{[a]}$, so that $[a + \Delta q] \in I_{[a]}$.

As far as the condition (6.8) is valid, the linear term in the expansion of the action is negligible. When Eq. (6.8) is violated the linear term leads to fast oscillations of the exponential of the action. Then $|U_{[a]}|$ becomes small (much less than the maximum) owing to destructive interference. The condition (6.8) is a formulation of the AUP in the case of monitoring the coordinate.

Monitoring an arbitrary observable This may be generalized to the case of monitoring an arbitrary observable $A(p, q, t)$ depending on momenta, coordinates, and time. Applying the phase-space form of the path integral we have for a partial propagator in this case

$$U_{[a]} = \int_{J_{[a]}} d[p,q] \exp\left\{ \frac{i}{\hbar} \int_0^T dt\, [p\dot{q} - H(p,q,t)] \right\}. \tag{6.9}$$

Here the set $J_{[a]}$ includes the paths $[p,q]$, giving the function

$$A(t) = A(p(t), q(t), t)$$

close to $[a]$ in the sense that $|A(t) - a(t)| \leq \Delta a(t)$.

An argument analogous to the above leads to an inequality analogous to Eq. (6.8), but with the functional derivatives in momenta and coordinates. This then leads to the following form of the AUP:

$$\max_{[\Delta p, \Delta q]} \left| \sum_i \int_0^T dt \left[\Delta p_i \left(\dot{q}_i - \frac{\partial H}{\partial p_i} \right) - \Delta q_i \left(\dot{p}_i + \frac{\partial H}{\partial q_i} \right) \right] \right| \lesssim \hbar, \tag{6.10}$$

where explicit expressions for the functional derivatives of the action $S[p,q]$ are used. The integrand here is taken for a path $[p,q]$ leading to $A(t) = a(t)$ (it will be shown at the end of the present section that the integrand does not depend on the choice of one of these paths). The maximum is calculated for the increments $[\Delta p, \Delta q]$ not bringing the path out of $J_{[a]}$, so that $A(t)$ corresponding to $[p', q'] = [p + \Delta p, q + \Delta q]$ differs from $a(t)$ less than by $\Delta a(t)$:

$$A(p(t), q(t), t) = a(t), \quad |A(p'(t), q'(t), t) - a(t)| \lesssim \Delta a. \tag{6.11}$$

Denoting $A(p'(t), q'(t), t) - a(t) = \Delta A(t)$, the condition (6.10) for $[a]$ being probable may be rewritten in the form

$$\max_{|\Delta A(t)| \leq \Delta a(t)} \left| \sum_i \int_0^T dt \left[\Delta p_i \left(\dot{q}_i - \frac{\partial H}{\partial p_i} \right) - \Delta q_i \left(\dot{p}_i + \frac{\partial H}{\partial q_i} \right) \right] \right| \lesssim \hbar. \tag{6.12}$$

It is important for the next argument that the integrand here is linear in Δp, Δq. Because of this linearity the condition (6.12) may be satisfied only if the factors multiplying Δp, Δq in the integrand are proportional to the components of the gradient of the function $A(p,q,t)$. Otherwise the integral may be made arbitrarily large by the corresponding choice of Δp, Δq even under the restriction $|\Delta A| \leq \Delta a$.

According to this, for Eq. (6.12) to be valid it is necessary that the following relations take place:

$$\dot{q}_i - \frac{\partial H}{\partial p_i} = -\delta F(t) \frac{\partial A(p,q,t)}{\partial p_i}, \quad \dot{p}_i + \frac{\partial H}{\partial q_i} = \delta F(t) \frac{\partial A(p,q,t)}{\partial q_i}. \tag{6.13}$$

Then the integrand in Eq. (6.12) contains the differential of the function $A(p,q,t)$. This differential is approximately equal to the increment ΔA. This gives for Eq. (6.12) the form

$$\max_{|\Delta A(t)|\leq \Delta a(t)} \left| \int_0^T dt\, \delta F(t) \Delta A(t) \right| \lesssim \hbar,$$

which is evidently equivalent to the following:

$$\int_0^T dt\, |\delta F(t)|\, \Delta a(t) \lesssim \hbar. \tag{6.14}$$

The inequalities (6.12) and (6.14) represent two forms of the AUP in the case of monitoring an arbitrary observable. The simplest form of the AUP is given by Eq. (6.14). It gives the restriction for the function $\delta F(t)$ determining the deviation of the measured system from the classical behavior. In the next section we shall comment on the physical meaning of the function δF.

It is clear from Eq. (6.13) that the integral in the l.h.s. of Eq. (6.10) does not depend on the choice of one of the paths $[p,q]$ satisfying Eq. (6.11) as has been assumed in the remark after Eq. (6.10). Indeed, if we go over from one of these paths $[p,q]$ to the other (close) $[p',q']$ the function $A(t)$ will not change. This means that the relation

$$\sum_i \left[\frac{\partial A(p,q,t)}{\partial p_i}(p_i' - p_i) + \frac{\partial A(p,q,t)}{\partial q_i}(q_i' - q_i) \right] = 0$$

holds. Taking Eq. (6.13) into account one sees that the integrand in the l.h.s. of Eq. (6.10) does not depend on whether the increments Δp, Δq are counted from $[p,q]$ or $[p',q']$.

6.3.3. QUANTUM MEASUREMENT NOISE AS A FICTITIOUS FORCE

In classical theory the measurement readout $[a]$ corresponds to the paths $[p,q]$ satisfying the classical Hamilton equations

$$\dot{q}_i - \frac{\partial H}{\partial p_i} = 0, \quad \dot{p}_i + \frac{\partial H}{\partial q_i} = 0.$$

According to Eq. (6.13), in the quantum theory of continuous measurements the readout $[a]$ also corresponds to the paths $[p,q]$ satisfying the classical Hamilton equations, but with an effective Hamiltonian

$$H_{\delta F}(p,q,t) = H(p,q,t) - \delta F(t)\, A(p,q,t),$$

where the function δF is restricted by the AUP (6.14).

The function δF may be interpreted as a generalization of a fictitious force. To prove this let us consider monitoring the coordinate q_j, i.e., take $A = q_j$. Let us use the notation δF_j in this case instead of the general notation δF. Then the additional term in the Hamiltonian has the form $-q_j \delta F_j(t)$. The deviation of the measured system from its classical behavior may therefore be described as the influence of an additional (fictitious) force $\delta F_j(t)$ corresponding to the coordinate q_j.

In the case of an arbitrary observable $A(p, q, t)$ the function δF determining the deviation from the classical behavior is a generalization of the force. The influence of this generalized fictitious force is equivalent to the inclusion of the term $-A(p, q, t)\, \delta F$ in the Hamiltonian.

Consider as a simple example monitoring the coordinate q in a harmonic oscillator with the Hamiltonian

$$H = \frac{p^2}{2m} + \frac{m\omega^2 q^2}{2} - q\, F(t).$$

Then $A = q$ and the additional term in the Hamiltonian has the form $-q\, \delta F(t)$, with $\delta F(t)$ satisfying the restriction (6.14). The paths corresponding to the measurement readout $[a]$ satisfy the equations (6.13), having in our case the form

$$\dot{q} - \frac{1}{m} p = 0, \quad \dot{p} + m\omega^2 q - F(t) = \delta F(t).$$

The curve $[a]$, according to the general recipe, may be obtained by substituting $p(t)$, $q(t)$ satisfying Eq. (6.13) in $A(t) = A(p(t), q(t), t)$. This gives in our case the function $a(t)$ satisfying

$$m\ddot{a} + m\omega^2 a = F(t) + \delta F(t). \tag{6.15}$$

Thus monitoring the coordinate of an oscillator gives a readout $[a]$ which looks as if the oscillator is classical but is driven by an additional force $\delta F(t)$ (in addition to the original force $F(t)$). This is a fictitious force describing the influence of the measurement process on the measured system. The value of this fictitious force is restricted by Eq. (6.14).

In the case considered the behavior of the measured observable is modified by the measurement (this is expressed by the r.h.s. of Eq. (6.15)). In the next section we shall consider the situation where the measurement, which has back action on the measured system, but nevertheless does not modify the behavior of the measured observable, so that the measurement readout $[a]$ satisfies the same equation as the observable A of the system without any measurement. The class of such measurements (observables) proves to coincide with the class of the so called quantum non-demolition (QND) measurements (observables) (BVK77; CTD$^+$80; Unr79).

6.3.4. QND OBSERVABLES

We see from Eqs. (6.13, 6.14) that for a small Δa (a precise measurement) the deviation of the system from the classical behavior is large. However, this deviation has the form specified by Eq. (6.13), and this provides the possibility for some observables to maintain classical behavior even in the course of the quantum measurement.We shall analyze this situation and prove that it arises when the observable undergoing monitoring is a quantum non-demolition observable.

Consider first the simple case (call it QND-1) when the classical equation for the observable A is of the form

$$\dot{A} = f(A, t).$$

The time derivative in this equation is owing to the Hamilton equations. Therefore it may be rewritten as

$$\frac{\partial A(p,q,t)}{\partial t} + \{H, A\}_{\mathrm{PB}} = f(A,t) \tag{6.16}$$

where

$$\{H, A\}_{\mathrm{PB}} = \sum_i \left(\frac{\partial H}{\partial p_i}\frac{\partial A}{\partial q_i} - \frac{\partial H}{\partial q_i}\frac{\partial A}{\partial p_i} \right)$$

is the Poisson bracket.

It is easy to demonstrate that the classical equation

$$\dot{a} = f(a, t)$$

for the measurement readout $[a]$ is a consequence of Eqs. (6.13, 6.16). It is enough for this to: (1) substitute $a(t) = A(p(t), q(t), t)$ in this equation; (2) express $\dot{a}$ through the partial derivatives of $A(p,q,t)$ and $\dot{p}$, $\dot{q}$; (3) substitute the expressions from Eq. (6.13) for $\dot{p}$, $\dot{q}$; and (4) use the classical equation of motion (6.16).

The observable A, which has a classical equation of the form $\dot{A} = f(A,t)$, is a QND observable. We have shown above that the monitoring of such an observable leads to the readout $[a]$ satisfying the same equation.

Consider now the more complicated case of an observable satisfying the second-order equation

$$\ddot{A} = f(A, \dot{A}, t).$$

This equation is equivalent to

$$A^{(2)} = f(A^{(0)}, A^{(1)}, t), \tag{6.17}$$

where we have denoted

$$A^{(0)} = A(p,q,t), \quad A^{(k+1)} = \frac{\partial A^{(k)}}{\partial t} + \{H, A^{(k)}\}_{\mathrm{PB}}, \quad k = 1,2,\ldots.$$

Applying a procedure analogous to that above we can show straightforwardly that the same classical equation for the measurement readout is satisfied,

$$\ddot{a} = f(a, \dot{a}, t),$$

provided that the equation (6.17) and the additional condition

$$\{A^{(0)}, A^{(1)}\}_{\mathrm{PB}} = 0$$

are valid. The observable A satisfying these conditions may be called a QND-2 type of observable.

Analogously the case of an observable of the QND-n type can be considered. Let the conditions

$$\begin{aligned} A^{(n)} &= f(A^{(0)}, A^{(1)}, A^{(2)}, \ldots, A^{(n-1)}, t), \\ \{A^{(0)}, A^{(k)}\}_{\mathrm{PB}} &= 0, \quad k \leq n-1, \end{aligned} \tag{6.18}$$

be satisfied. In this case the classical equation for time dependence of the observable A has the form

$$A^{(n)}(t) = f(A, \dot{A}, \ddot{A}, \ldots, A^{(n-1)}, t).$$

It may be shown that the continuous measurement of such an observable leads to the readout $[a]$ satisfying the same classical equation of motion

$$a^{(n)}(t) = f(a, \dot{a}, \ddot{a}, \ldots, a^{(n-1)}, t).$$

If we are not interested in separating all QND variables in the classes QND-n, then the condition for the observable $A(p,q,t)$ to belong to one of these classes may be written in the form

$$\{A^{(k)}, A^{(l)}\}_{\mathrm{PB}} = 0, \quad \text{for arbitrary} \quad k, l.$$

The same may be written in terms of the Heisenberg operators

$$\hat{A}_H(t) = e^{\frac{i}{\hbar}\int_0^t \hat{H}\,dt}\, A(\hat{p}, \hat{q}, t)\, e^{-\frac{i}{\hbar}\int_0^t \hat{H}\,dt}$$

as a series of conditions

$$\left[\hat{A}_H^{(k)}(t), \hat{A}_H^{(l)}(t)\right] = 0 \quad \text{for arbitrary} \quad k, l.$$

Finally, the same may be rewritten as the requirement for the Heisenberg operators to commute at different times:

$$\left[\hat{A}_H(t), \hat{A}_H(t')\right] = 0, \quad \text{for} \quad t, t' \in [0, T].$$

6.3.5. UNCERTAINTY RELATIONS FOR CONTINUOUS MEASUREMENTS

The inequality (6.12) implies some uncertainty relations for continuous quantum measurements.

To make further consideration more transparent, let us introduce the notation $F_i = \partial H/\partial q_i$ for the force corresponding to the coordinate q_i and assume that the kinematic term of the Hamiltonian has its usual form so that $\partial H/\partial p_i = p_i/m$. Then the integrand in Eq. (6.12) has the form $\sum_i \delta\sigma_i$ where

$$\delta\sigma_i(t) = \Delta p_i(t)\, \delta q_i(t) - \Delta q_i(t)\, \delta p_i(t)$$

with

$$\delta q_i = dq_i - \frac{1}{m} p_i dt, \quad \delta p_i = dp_i - F_i dt.$$

The entity $\delta\sigma_\imath$ has a simple geometric meaning (see Fig. 6.1). This is

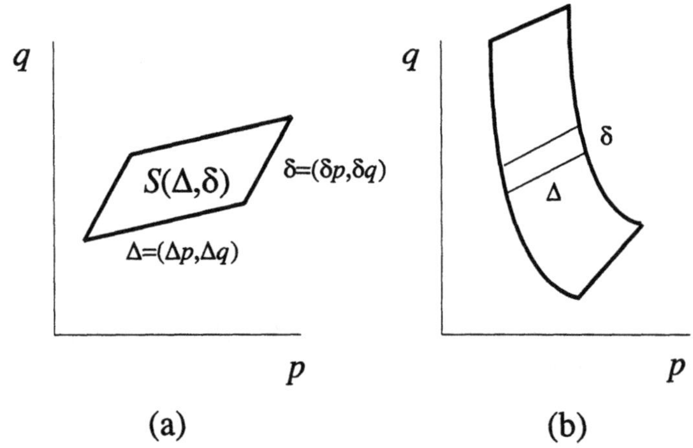

Figure 6.1. Uncertainties in a continuous measurement expressed by an area in the phase space: the elementary area defined by the measurement uncertainty and infinitesimal interval of evolution (*a*), and the area of the strip corresponding to a finite interval of evolution (*b*). For the corresponding measurement readout to have high probability the area of the strip must be less than $\hbar$.

(up to the sign) an area of the parallelogram in the phase space (p_i, q_i) constructed with the help of the vectors $(\Delta p_i, \Delta q_i)$ and $(\delta p_i, \delta q_i)$. The first of these vectors characterizes the resolution of the continuous measurement, and the second vector shows how much the momentum and coordinate deviate from their classical law of evolution.

According to Eq. (6.12),

$$\max_{|\Delta A(t)|\leq \Delta a(t)} \left| \sum_i \int \delta\sigma_i \right| \lesssim \hbar,$$

where the integration is performed along the trajectory in the phase space. This inequality must be valid for any measurement readout $[a]$ emerging with high probability. The l.h.s. of this inequality may be arbitrarily small and even zero (for the readout coinciding with the classical trajectory $[a_{\text{class}}]$. However, for some of the probable readouts $[a]$ and for some increments $[\Delta p, \Delta q]$ the l.h.s. of the inequality achieves its maximum possible value $\hbar$. Therefore one may conclude that

$$|\sum_i \int \delta\sigma_i| \sim \hbar$$

for some 'typical' probable readouts and some 'typical' Feynman trajectories contributing to the evolution of the measured system. The time integral here gives an area of the 'band' determined by the vectors $(\Delta p_i(t), \Delta q_i(t))$ in the transversal direction and by the vectors $(\delta p_i(t), \delta q_i(t))$ in the longitudinal direction.

For a one-dimensional system this gives

$$\left| \int (\Delta p\, \delta q - \Delta q\, \delta p) \right| \sim \hbar,$$

where Δp, Δq specify the resolution of the measurement and δp, δq the deviation of the measured system from the classical behavior. If the duration of the measurement is short we have

$$|\Delta p\, \delta q - \Delta q\, \delta p| \sim \hbar. \tag{6.19}$$

The classical uncertainty relation for the momentum and time (LL87) may be derived (in a slightly different form) as a special case of Eq. (6.19). To this end let us consider measuring the momentum p during a short time Δt with the resolution Δp. Suppose that the coordinate q is measured, in the process of the continuous measurement, very roughly, i.e., $\Delta q \to \infty$. Then Eq. (6.19) may be fulfilled only for $\delta p = 0$, i.e., $p = \text{const}$ (in the case of the null force, $F = 0$). Therefore the momentum is not changed during the measurement. The uncertainty relation (6.19) now reads

$$\Delta p\, |\delta q| = \Delta p \left| (q'' - q') - \frac{p}{m} \Delta t \right| \sim \hbar.$$

Here p denotes a result of the measurement of the momentum, and q', q'' are 'boundary conditions' i.e., the values of the coordinate before and after the continuous measurement (for example, these values may be obtained by 'instantaneous' (i.e., very short) measurements of the coordinate at the instants $t = 0$ and $t = T$).

The interpretation of this inequality is evident. As a result of the continuous measurement of the momentum the increment of the coordinate $q'' - q'$ may differ by the amount $\delta q = \hbar/\Delta p$ from its classically predicted value $p\Delta t/m$.

In other words, owing to measuring the momentum, the velocity $(q'' - q')/\Delta t$ may vary by a value of the order of $\hbar/\Delta p \Delta t$ from its classical value p/m. This may be compared with the uncertainty relation derived by the analysis of a gedanken experiment (LL87):

$$\Delta p \, |v'' - v'| \Delta t \sim \hbar,$$

where v' and v'' are the velocities before and after the measurement.

6.4. Comments

The RPI (quantum corridor) approach as an information approach to quantum measurements has been discussed in (Men96b; Men98a). The role of continuous measurements in the mathematical foundations of path integrals was discussed in (Men92c). The action uncertainty principle was formulated and proved for linear systems in (Men91; Men92a; Men92b). In (Men96a) it was formulated in a number of different forms, analyzed and proved for nonlinear as well as linear systems. The group-theoretical analysis of continuous measurements is taken from the book (Men93).

Part III

Continuous Quantum Measurements: Applications

Chapter 7

The Monitoring of Energy

Theory of continuous measurements developed in the preceding chapters is applied in the present chapter to a simple but realistic example of a fuzzy continuous measurement of energy in a two-level system. How such a measurement may be realized will be shown in Chapter 8. In the present chapter we apply the phenomenological restricted-path-integral approach to explore some specific features of this measurement. An advantage of this measurement which will be demonstrated is that it allows one to monitor a quantum transition between the levels.

So far level transitions (induced for example by action of the laser radiation upon an atom) were observed experimentally as instantaneous events. Their time evolution (for example, Rabi oscillations) were experimentally verified only in the statistics of many transitions with the same initial conditions but with energy measured at various instants. It seemed impossible to explore experimentally the time evolution of a transition in a *single* quantum process. This is actually impossible if the measurement of energy is sharp (projective). The reason is, of course, clear: the sharp measurement interrupts the process of transition, projecting the system onto one of these levels. A consequence of this is the well known quantum Zeno effect (Sect. 4.1).

However, a fuzzy continuous measurement of energy (realized for example as a series of fuzzy instantaneous observations, as argued in Chapter 8) allows the monitoring of the transition experimentally, the visualization of the transition of a single quantum system (atom). This is shown in Sect. 7.3. Unavoidable quantum restrictions on the monitoring are derived. Soft (fuzzy) continuous quantum measurements prove to supply a principally novel instrument for investigating quantum processes.

Sect. 7.2 is devoted to the fuzzy continuous measurement of energy in a free (not driven) multi-level system. This is interesting as a simple and transparent example of continuous measurement. Moreover, this example shows that continuous measurement may serve as a model of the process of decoherence as it has been defined in Sect. 2.1 (comp. Sect. 2.4): the physical process gradually improves the resolution of the levels and finally leads to complete resolution, i.e., to projective measurement of an energy level.

7.1. The monitoring of energy

The phenomenological theory for the monitoring of an arbitrary observable A has been presented in Sect. 5.1. The theory starts with the restricted path integral representation for partial propagators (corresponding to various measurement readouts) and then, for the case of monitoring, is reduced to the Schrödinger equation with a complex Hamiltonian, depending on the readout.

Now we turn to the concrete case of the measurement of energy. Let the Hamiltonian H of the system have the form $H = H_0 + V$, where H_0 is the Hamiltonian of the 'free' multi-level system and V is a potential describing an external influence leading to transitions between levels if no measurement is applied. By measurement of the energy we shall mean the measurement of the observable $A = H_0$ leading to the readout $[E] = \{E(t')|0 \leq t' \leq T\}$.

The weight factor (5.11) with $A = H_0$ and $[a] = [E]$ can be written in the form:[1]

$$\begin{aligned} w_{[E]}[p,q] &= -\kappa \int_0^T \Big(H_0(p(t),q(t),t) - E(t)\Big)^2 dt \\ &= \exp\left[-\frac{\langle (H_0 - E)^2\rangle_T}{\Delta E_T^2}\right], \end{aligned} \tag{7.1}$$

where we have made use of the notation

$$\langle (H_0 - E)^2\rangle_T = \frac{1}{T}\int_0^T \Big(H_0(p(t),q(t),t) - E(t)\Big)^2 dt \tag{7.2}$$

and introduced the parameter

$$\Delta E_T = \frac{1}{\sqrt{\kappa T}} \tag{7.3}$$

characterizing the fuzziness of the measurement in a more transparent way than κ.

The weight factor (7.1) guarantees that only those Feynman paths $[p,q]$ are taken into account for which the time dependence of the observable H_0 is represented by a curve close to the curve $[E]$. The measure of the deflection $\langle (H_0 - E)^2\rangle_T$ is, according to (7.1), less than or of the order of ΔE_T^2. This means that the *resolution* of the measuring device described by our formulas is equal to ΔE_T.

[1]It is only for simplicity that we assume κ to be constant. It may depend upon time, and then it must be included in the integrand.

The parameter κ characterizes properties of the measurement device and therefore does not depend upon the duration of the measurement process. Therefore κ does not depend upon T. Itthen follows from Eq. (7.3) that the resolution of the continuous fuzzy measurement is improved when the duration of the measurement increases. This important property may be understood if one replaces the continuous fuzzy measurement by a sequence of instantaneous fuzzy measurements with Gaussian weight functionals. It is plausible that the repetition of fuzzy measurements improves the resulting sharpness. Therefore in the case of a constant frequency of repetitions (corresponding to constant κ) the resolution is improved if the series of repetitions is longer.

The measurement of energy is now worked out as follows (Sect. 5.1): Let the measured system be initially in a pure state $|\psi_0\rangle$. Then for a particular given readout $[E]$ its further evolution is obtained by solving the corresponding Schrödinger equation (comp. (5.15) with $A = H_0$ and $[a] = [E]$):

$$\frac{\partial}{\partial t}|\psi_t\rangle = \left(-\frac{i}{\hbar}H - \kappa\left(H_0 - E(t)\right)^2\right)|\psi_t\rangle. \tag{7.4}$$

The norm of the final state vector $|\psi_T\rangle$ then gives according to the general formula (5.16), the probability density for the measurement readout $[E]$:

$$p[E] = \langle\psi_T|\psi_T\rangle. \tag{7.5}$$

This together with the resulting state $|\psi_T\rangle$ is what can be known about the measurement.

Expanding the state $|\psi_t\rangle$ in the basis

$$|\varphi_n(t)\rangle = e^{-iE_n t/\hbar}|n\rangle \tag{7.6}$$

of time-dependent eigenstates of H_0, we have the following system of equations for the coefficients of the expansion $|\psi_t\rangle = \sum C_n(t)|\varphi_n(t)\rangle$:

$$\dot{C}_n = -\kappa(E_n - E(t))^2\, C_n - \frac{i}{\hbar}\sum_{n'}\langle\varphi_n|V|\varphi_{n'}\rangle C_{n'}. \tag{7.7}$$

The probability density of the measurement's readouts is given by

$$p[E] = \sum_n |C_n(T)|^2. \tag{7.8}$$

7.2. Free multi-level system

We shall consider first a 'free' multi-level system with the Hamiltonian H_0. 'Free' means vanishing V, so there is no driving field and no transitions between levels occur. We shall see in this simple case that a continuous fuzzy measurement of energy may serve as a model for decoherence leading to the resolution of the energy levels.

Because of $V = 0$, Eq. (7.7) has the simple solution

$$\begin{aligned} C_n(T) &= C_n(0) \exp\left[-\kappa \int_0^T dt\,(E_n - E(t))^2\right] \\ &= C_n(0) \exp\left[-\frac{\langle (E_n - E)^2\rangle_T}{\Delta E_T^2}\right], \end{aligned} \tag{7.9}$$

where $\langle\,\rangle_T$ is a time average defined in (7.2) and ΔE_T is defined by Eq. (7.3).

7.2.1. PROBABILITY DISTRIBUTION

A probabilistic analysis may be based on Eqs. (7.5, 7.8)). In our case the probability density

$$\begin{aligned} p[E] &= \sum_n |C_n(T)|^2 \\ &= \sum_n |C_n(0)|^2 \exp\left[-2\frac{\langle (E_n - E)^2\rangle_T}{\Delta E_T^2}\right] \end{aligned} \tag{7.10}$$

is exponentially small when the temporal mean squared deflection $\langle (E_n - E)^2\rangle_T$ of the function $[E]$ from each level E_n is more than ΔE_T^2. In contrast, the probability density has a local maximum near the function $[E]$ for which $\langle (E_n - E)^2\rangle_T$ is much smaller than ΔE_T^2 for a level E_n corresponding to nonzero $|C_n(0)|^2$.

Let us now turn to the probability itself. The generalized unitarity in the case of the energy measurement has the form

$$\int d[E]\, U_{[E]}^\dagger U_{[E]} = 1.$$

Taking matrix elements of this equality in the energy representation we have an equivalent condition

$$\int d[E] \exp\left[-2\frac{\langle (E - E_n)^2\rangle_T}{\Delta E_T^2}\right] = 1 \quad (\text{for all } n). \tag{7.11}$$

The path integral here is of Gaussian type and can be evaluated (see Appendix A3). It may be shown that the (generalized) unitarity condition is satisfied if the functional measure $d[E]$ is chosen to be

$$d[E] = \prod_t \sqrt{2\kappa dt/\pi}\, dE(t).$$

Because of the exponential in the integrand, only paths close enough to the level E_n contribute in (7.11); namely, the integral which is analogous but restricted to the subset of readouts

$$\mathcal{A}_n = \{[E] \mid \langle (E - E_n)^2 \rangle_T \lesssim \Delta E_T^2\}, \tag{7.12}$$

is close to unity:

$$\int_{\mathcal{A}_n} d[E] \exp\left[-2\frac{\langle (E-E_n)^2\rangle_T}{\Delta E_T^2}\right] \approx 1. \tag{7.13}$$

The probability that the measurement output $[E]$ belongs to an arbitrary set $\mathcal{A}$ is according to the general formula (5.17), and with the help of Eq. (7.10)

$$\begin{aligned} \mathrm{Prob}([E] \in \mathcal{A}) &= \int_{\mathcal{A}} d[E] \,\mathrm{Tr}\left(U_{[E]}\, \rho_0\, U_{[E]}^{\dagger}\right) \\ &= \sum_n p_n(\mathcal{A})\, |C_n(0)|^2 \end{aligned} \tag{7.14}$$

with

$$p_n(\mathcal{A}) = \int_{\mathcal{A}} d[E] \exp\left[-2\frac{\langle (E-E_n)^2\rangle_T}{\Delta E_T^2}\right]. \tag{7.15}$$

Comparing this with Eq. (7.13) we see that $p_n = 1$ if $\mathcal{A}_n \subset \mathcal{A}$ (or the intersection of $\mathcal{A}_n$ with $\mathcal{A}$ is only slightly less than $\mathcal{A}_n$). If, on the contrary, the set $\mathcal{A}$ does not intersect with $\mathcal{A}_n$ (or their intersection is small compared with $\mathcal{A}_n$) then $p_n = 0$ (or close to zero).

7.2.2. DIFFERENT REGIMES OF MEASUREMENT

As has already been said, the measurement resolution ΔE_T decreases for increasing duration T of the measurement. To define different regimes of the measurement this resolution should be compared with the characteristic difference $\Delta E \sim E_{n+1} - E_n$ between energy levels.[2] It is more convenient, however, to go over to the corresponding time parameters. Let T_{lr} be the duration at which the measurement resolution ΔE_T achieves the value ΔE,

$$T_{\mathrm{lr}} = \frac{1}{\kappa \Delta E^2}. \tag{7.16}$$

Then ΔE_T is larger or smaller than ΔE depending on whether T is smaller or larger in comparison with T_{lr}.

[2]If the difference between levels substantially depends upon the number n of the level, one can draw corresponding conclusions concerning the measurement in a certain energy band referring to the value ΔE which is typical for this band.

From Eq. (7.9) we have for the coefficients C_n at time $t = T_{\text{lr}}$

$$C_n(T_{\text{lr}}) = C_n(0) \exp\left[-\frac{\langle (E_n - E)^2 \rangle_{T_{\text{lr}}}}{\Delta E^2}\right]. \tag{7.17}$$

Therefore the coefficient $C_n(t)$ decreases essentially during the time T_{lr} if the deflection of $E(t)$ from E_n is of the order or greater than ΔE (this could not happen for much shorter measurement). This is why the parameter T_{lr} may be called the *level resolution time.*

Let us now consider different durations T and the corresponding regimes of measurement. In the case $T \gg T_{\text{lr}}$ we have $\Delta E_T \ll \Delta E$. Now the probability density (7.10) is not negligible only if for most of the measurement's duration

$$|E - E_n| \lesssim \Delta E_T \ll \Delta E \tag{7.18}$$

for some n. The corresponding set of readouts $\mathcal{A}_n$ determined by Eq. (7.12) is in this case a very narrow band close to the constant curve $E(t) \equiv E_n$. The width of this band is much less than ΔE, so no other level $E_{n'}$ with $n' \neq n$ lies inside $\mathcal{A}_n$. Taking the set of paths $\mathcal{A}$ in (7.14) in such a way that it includes $\mathcal{A}_n$ but does not intersect with any other $\mathcal{A}_{n'}$, $n' \neq n$, we have $p_n = 1$ and $p_{n'} = 0$ for $n' \neq n$. Therefore it follows from (7.14) that

$$\text{Prob}(\mathcal{A}) = |C_n(0)|^2. \tag{7.19}$$

Therefore in the case $T \gg T_{\text{lr}}$ the measurement readout $[E]$ will be an almost constant curve close to one of the levels E_n, and the probability of it being close to a specified E_n is $|C_n(0)|^2$. We see that the continuous fuzzy measurement, if it lasts longer than T_{lr}, distinguishes (resolves) between the levels. This justifies once again the name *level resolution time* for the parameter T_{lr}. To find the state of the system after the measurement in the case considered ($T \gg T_{\text{lr}}$) we have to use Eq. (7.9) with the function $E(t)$ satisfying the restriction (7.18). We obtain

$$C_n(T) = C_n(0), \quad C_{n'}(T) = 0, \quad n' \neq n. \tag{7.20}$$

Thus, after a continuous fuzzy measurement of duration longer than T_{lr} the situation is similar to that in the conventional description of the energy measurement as given in the von Neumann postulate (Sect. 2.3): (1) The only possible measurement readouts $[E]$ are those close to one of the energy levels E_n; (2) the probability of $[E]$ being close to a particular E_n is $|C_n(0)|^2$; and (3) in the case of $[E]$ being close to E_n the system turns out to be in the energy eigenstate $|\varphi_n\rangle$ after the measurement.

The physical process leading to the determination of an energy eigenvalue (or of the eigenvalue of any other observable with discrete spectrum) has been investigated in the framework of different models (JZ85;

Zur82) and was called the process of decoherence (Sect. 2.1). We see now that the continuous fuzzy measurement of energy fixes the energy level during T_{lr}. Therefore the result of the continuous measurement is just the same as the result of decoherence.

Below, considering brief measurement, $T \ll T_{\text{lr}}$, we shall see that the *continuous measurement supplies a model of the whole process of gradual decoherence* leading to the determination of the specific energy eigenvalue.

The characteristic mark of decoherence is the dying out of the off-diagonal matrix elements of the density matrix. It is easily seen that it is this which really happens at the time T_{lr} because owing to (7.17) all products $C_i C_j^*$ with $i \neq j$ become exponentially small for $T \gg T_{\text{lr}}$. The time period during which non-diagonal elements die out, is usually called the decoherence time. We call it the level resolution time to be more concrete, and because we consider other regimes of measurement too, where the aim is not necessarily level resolution.

In the context of the above arguments a continuous measurement performed during the time $T \ll T_{\text{lr}}$ might be considered as not yet finished. However, this is valid only if the aim of the measurement is to distinguish between different levels. Continuous fuzzy measurement may have as a goal the estimation of the energy of the system with a resolution worse than ΔE. Then a period of measurement shorter than T_{lr} makes sense too. Let us turn to this regime.

We consider the case $T \ll T_{\text{lr}}$. This is equivalent to $\Delta E_T \gg \Delta E$, so that the set of paths $\mathcal{A}_n$ determined by Eq. (7.12) includes functions $[E]$ in a very wide energy band around the constant curve $E \equiv E_n$. The width of this band is ΔE_T, so that it includes many other energy levels. Vice versa, if we define the set of readouts $\mathcal{A}$ to contain all functions $[E]$ in the energy band of width ΔE_T between E_{min} and E_{max}, then many energy levels E_n, $n_1 \leq n \leq n_2$ lie inside this band. For each of these levels intersection of $\mathcal{A}_n$ and $\mathcal{A}$ makes a significant part of $\mathcal{A}_n$, and therefore $p_n(\mathcal{A}) \sim 1$ for $n_1 \leq n \leq n_2$. On the other hand, for n outside the interval $[n_1, n_2]$ intersection between $\mathcal{A}_n$ and $\mathcal{A}$ is small so that $p_n(\mathcal{A})$ is small. Making use of Eq. (7.14) we have approximately

$$\text{Prob}(\mathcal{A}) = \sum_{n=n_1}^{n_2} |C_n(0)|^2. \tag{7.21}$$

This regime of measurement does not distinguish between the energy levels so the character of the measurement in this regime is essentially the same as in the case of a continuous spectrum.

Summarising, the continuous measurement of the duration T results in energy levels surviving in an energy band of the width ΔE_T with all the rest dying out. Where the center of this band lies is a random event, the probability of a band being equal to the sum of the values $|C_n(0)|^2$ for the levels in the band. When the duration T grows the energy band becomes more narrow and finally, for a duration of the order of T_{lr}, it becomes of the order of ΔE. After this only one energy level survives.

This picture is nothing other than a portrayal of the gradual decoherence. The phenomenological theory of continuous measurement simultaneously turns out to be a *theory of gradual decoherence.* This statement will be supported by the considerations in Chapter 8. There a realization of the continuous measurement of energy by a long series of short weak interactions will be proposed. It is clear that such a series of interactions may serve as a model for the interaction of the system with its environment. The phenomenological RPI or complex-Hamiltonian approach describes the same, but without any concrete model.

From another point of view, a measurement much shorter than T_{lr} may be called (with respect to the given energy levels) fuzzy (unsharp). Its influence on the system is quite weak, so the measurement may also be called soft. If, in contrast, the measurement is much longer than T_{lr} it is sharp or projective. Such a measurement catches the system on some level (one of those for which $|C_n(0)|^2$ is nonzero) and keeps it at this level.

So far we have discussed only a free system, $V = 0$, but we shall see that the same regimes of measurement are also characteristic for a driven system $V \neq 0$.

7.3. Monitoring of a level transition

Let us consider now the question of whether the time development of the state can be made visible by the measurement readout (AM97; AMN97; AM98). We shall consider a system under an influence which may cause a transition between levels. If energy is then continuously measured in this system we may hope that this will give us (at least approximately) a picture of a gradual transition from one level to another.

This problem is much harder than the preceding one, so we shall restrict ourselves by the case of a two-level system under the influence of a periodic force in resonance with the energy difference, $\hbar\omega = \Delta E$. The Schrödinger equation with effective Hamiltonian (7.7) then takes the form

$$\begin{aligned}\dot{C}_1 &= -iv(t)C_2 - \kappa(E_1 - E(t))^2 C_1, \\ \dot{C}_2 &= -iv(t)C_1 - \kappa(E_2 - E(t))^2 C_2\end{aligned} \tag{7.22}$$

where v is proportional to the matrix elements of the potential,

$$\hbar v = \langle\varphi_1|V|\varphi_2\rangle = \langle\varphi_2|V|\varphi_1\rangle^* \tag{7.23}$$

which are constant or alter slowly compared to the frequency ω.

Let us describe first the case when there is no continuous measurement ($\kappa = 0$) and the wave function is normalized. Then owing to the driving field V the state of the system undergoes Rabi oscillations between the eigenstates with Rabi period $T_R = \pi/v$. This means that for ($\kappa = 0$) we have $C_n(t) = R_n(t)$, where

$$\begin{aligned} R_1(t) &= C_1(0)\cos vt - iC_2(0)\sin vt, \\ R_2(t) &= C_2(0)\cos vt - iC_1(0)\sin vt. \end{aligned} \tag{7.24}$$

Our task will be to find out what will be the behavior of the system if $\kappa \neq 0$, i.e., if the continuous measurement is performed, and what will be the relation between the measurement readout $[E]$ and the behavior of the system.

First of all it is seen from Eq. (7.22) that the behavior of the system will be the same (at least during a quite short time interval) if κ is small. In this case the measurement is very weak or fuzzy, so it influences the system insignificantly. In this respect such a measurement may be called soft. In the opposite case of large κ (a sharp measurement) the measurement influences the system strongly. It may be expected that the system will be caught and frozen at one of the levels (the quantum Zeno effect, see Sect. 4.1). Such a measurement might be characterized as hard. We shall see that all this actually takes place, and will also explore the most interesting intermediate regime of measurement when κ is neither too large nor too small.

More concretely, let us consider a π-pulse of the resonance driving field in the interval $[T_1, T_2]$, where $0 \leq T_1 \leq T_2 \leq T$. This means that $T_2 - T_1 = T_R/2$ and v in the equation (7.22) vanishes outside this interval. If the system is on level 1 at the time T_1, i.e., $C_1(T_1) = 1$, $C_2(T_1) = 0$, then without the measurement the system will perform the level transition during the interval $[T_1, T_2]$ with certainty. The probability $|C_2(t)|^2$ of being at level 2 will gradually increase during this interval and will reach unity at the end of it. The system will stay at level 2 also after $t = T_2$ and therefore it will be with certainty at level 2 at $t = T$. We shall see how this process is modified if the continuous measurement of energy is performed.

A quantity by which in this case the fuzziness of this measurement can be characterized and which can replace κ in physical discussions is the *level resolution time* (7.16). If a continuous fuzzy measurement lasts much longer than T_{lr} it distinguishes (resolves) between the levels. Small T_{lr} represents quick level resolution because of small fuzziness, and therefore a strong Zeno-type influence of the measurement may be expected in this case. Such a measurement may be called *hard* and the corresponding regime the *Zeno regime*. The larger T_{lr} is, the weaker is the influence of the measurement on the atom. The Rabi oscillations may show up. A large T_{lr} therefore corresponds to a *soft* measurement (the *Rabi regime*). We now present some results of a numerical analysis of Eq. (7.22) for three regimes of measurement: the Zeno-type, the Rabi-type, and the intermediate regimes.

Assuming the influence of a continuous measurement of energy with a given fuzziness on a quantum system, we want to treat the following problems:

i) For a given readout $[E]$, what is the corresponding development of the state $|\Psi_t\rangle$ of the system as represented by the modulus $|c_n(t)|$ of the normalized coefficient $c_n(t) = C_n(t)/\sqrt{|C_1(t)|^2 + |C_2(t)|^2}$? We denote the whole curve $\mathcal{P}_2(t) = |c_n(t)|^2$ by $[\mathcal{P}_2]$. $\mathcal{P}_2(t)$ is the probability of finding the system in the state 2 if a projection measurement were to be made at the time t.

ii) In addition we want to know the probability density $p[E]$ of obtaining a particular readout $[E]$ which is simultaneously a probability density for the related state curve $[\mathcal{P}_2]$.

In the course of the numerical analysis random curves $E(t)$ were generated, for each of them Eq. (7.22) was solved, the norm of the solution at the final time $|C_1(T)|^2 + |C_2(T)|^2$ was calculated, and according to Eqs. (7.5, 7.8) interpreted as the probability density $p[E]$ that the curve $E(t)$ occurs as the measurement readout. The related state curve $\mathcal{P}_2(t)$ was worked out. Being determined as soon as $[E]$ is fitted, $[\mathcal{P}_2]$ carries the same probability density as $[E]$. For further analysis all curves $E(t)$ were smoothed with the time scale of the order of $T_2 - T_1$ to eliminate insignificant fast oscillations but conserve information about the transition. We discuss the results for the different regimes separately.

7.3.1. THE TWO LIMITING CASES

The results of the analysis for the two limiting choices of fuzziness are given in the density plots of Fig. 7.1. The smoothed curves $E(t)$ are presented in the upper diagrams and the corresponding curves $\mathcal{P}_2(t)$ in the lower diagrams. The degree of greynessi at each point of the diagram indicates the probability density that a curve passes through this point. The figure thus gives an idea of the limits in which a relatively probable measurement readout must lie.

The left part of Fig. 7.1 refers to a soft continuous measurement. It belongs to the Rabi regime, where the influence of the driving field dominates. Regarding the development of the state, the measurement only slightly influences the measured system, not preventing the level transition which occurs with high probability. But in this case the information about the behavior of the system which can be obtained from the readouts of the upper left diagram is vague because the band of probable measurement readouts is wide. The measurement readout $[E]$ may even remain close to

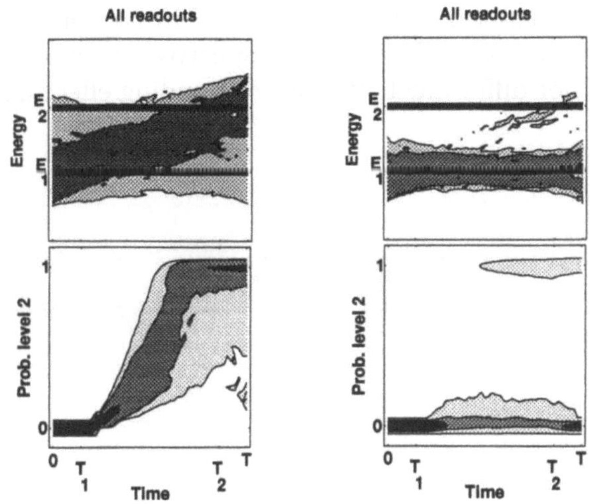

Figure 7.1. Density plots of measurement readouts $E(t)$ and the corresponding probability curves $\mathcal{P}_2(t) = |c_2(t)|^2$ characterizing the behavior of the state of the system in time for two values of fuzziness: a soft measurement with $4\pi T_{lr}/T_R = 10/3$ (left) and $4\pi T_{lr}/T_R = 2/3$ (right).

level E_1 (pointing out the absence of the transition), although the system will most probably transit to level E_2 just as without any influence of the measurement.

The right half of Fig. 7.1 refers to the opposite limit. It demonstrates the characteristic features of a hard measurement (the Zeno regime). With high probability the measurement readout $[E]$ shows that the system stays on level E_1, and the character of the curves $[\mathcal{P}_2]$ confirms this. The hard (Zeno-type) measurement influences the measured system very strongly, thus preventing with high probability the level transition. The information about the development of the system which can be read off from a measured curve $E(t)$ of the upper right diagram is much more precise (a narrow band of probable measurement readouts) and reliable. But the price for this is that the Rabi oscillation is completely suppressed.

7.3.2. RELIABILITY IN THE INTERMEDIATE REGIME

For the regime of measurement which is intermediate between Zeno and Rabi the results are also intermediate and much more interesting: The influence of the measurement on the system is in this case less than in the Zeno regime, and the information given by the measurement readout $E(t)$ about the system's behavior is more precise than in the Rabi regime.

The left pair of diagrams in Fig. 7.2 shows all readouts $E(t)$ and the related state curves $\mathcal{P}_2(t)$ in this intermediate regime. It can be seen that the curves $E(t)$ are separated at the final time T in two branches approaching the upper or lower energy level. The respective probabilities of obtaining these curves do not differ much. The corresponding effect appears for $\mathcal{P}_2(t)$.

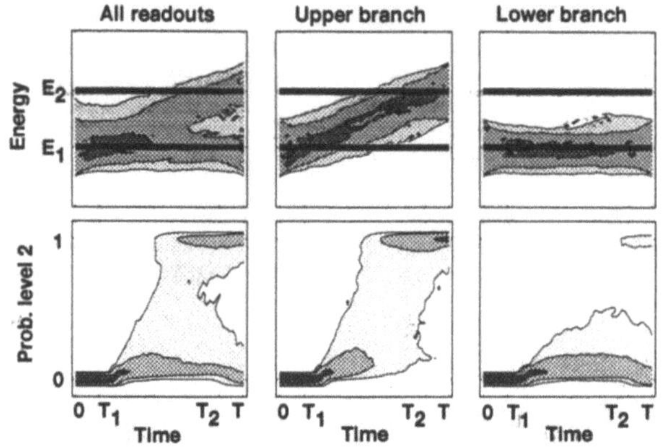

Figure 7.2. Continuous measurement of a moderate fuzziness ($4\pi T_{lr}/T_R = 4/3$). In the left pair of diagrams all measurement readouts $E(t)$ and curves $\mathcal{P}_2(t) = |c_2(t)|^2$ are shown. In the middle and right hand diagrams readouts are selected which point out that a transition to level 2 happens (middle) or does not happen (right). The respective curves $\mathcal{P}_2(t)$ confirm that the state behaves like this.

It is now interesting to compare the curves in more detail in separating the readouts into two classes. The middle diagrams of Fig. 7.2 refer to readouts pointing to a level transition and the diagrams on the right to its prevention. It can be read off directly that in this regime the information obtained from the readout $E(t)$ is quite reliable. This means that a readout $E(t)$ and the corresponding curve $\mathcal{P}_2(t)$ for the development of the state are with high probability in accord with each other. If, for example, $E(T)$ is close to level E_2, then the most probable behavior of $\mathcal{P}_2(t)$ is characteristic for the transition from level 1 to level 2. For $E(T)$ close to E_1 the system stays with high probability at level E_1 according to $\mathcal{P}_2(t)$. Note that there is an approximately equal probability for a transition to occur or not to occur.

The numerical analysis has therefore yielded the following important *result*: The visualization of an externally driven quantum transition by a continuous measurement of energy is possible if the fuzziness is chosen appropriately, namely to be of intermediate strength. In this case one may deduce the actual behavior $\mathcal{P}_2(t)$ of the measured system from the measurement's readout $E(t)$. However, the measurement (even in the intermediate regime) influences the system. As a result, the probability of the transition from level 1 to level 2 becomes less than unity. It is not possible to predict

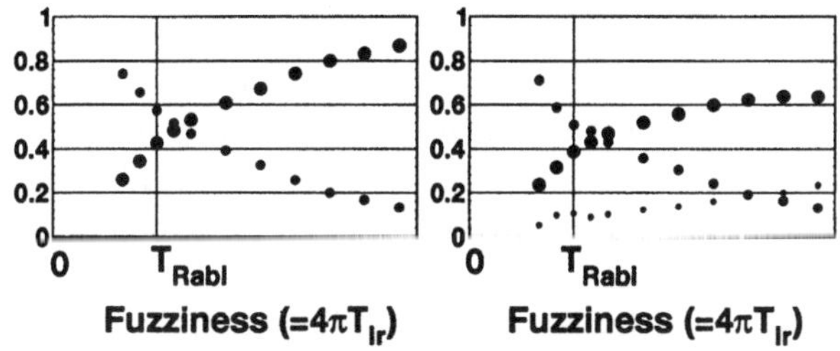

Figure 7.3. Probabilities that a transition to the upper level happens (big points) or does not happen (middle sized points) as functions of the fuzziness. In the left diagram these probabilities are read off from the curves $\mathcal{P}_2(t)$, describing the evolution of the state and in the right diagram from the energy readouts $E(t)$. Small points present the measurement noise.

a single transition with certainty; the transition may occur or not occur with probabilities of the same order. This quantum uncertainty is the price of the visualization of the quantum transition.

7.3.3. DEPENDENCE ON THE FUZZINESS

In Fig. 7.3 we present the effects discussed above from a different point of view and show how they change gradually with increasing fuzziness. For this purpose we renounce exhibiting the total curves $[E]$ and $[\mathcal{P}_2]$ as functions of time. Instead, for a given fuzziness represented by the quantity $4\pi T_{lr}$ we divide the curves into two classes, one class corresponding to a transition and another to the absence of a transition. The probability of a curve to belong to one of the two classes is worked out.

In the left hand diagram of Fig. 7.3 we refer to the curves $[\mathcal{P}_2]$ describing the time development of the state of the system. If this curve ends close to unity, in the sense of $\mathcal{P}_2(T) > 1/2$, we say that the transition has occurred. The big points in the left hand diagram represent the probability of this happening as a function of the fuzziness. For $\mathcal{P}_2(T) < 1/2$ the same is indicated by middle size points. The diagram shows how the probability for transitions to occur increases with increasing fuzziness. For the fuzziness $4\pi T_{lr} = (10/3)T_R$ at the right hand end of the diagram the probability for a transition is 88 % (the Rabi regime of measurement).

With regard to the energy readout it is assumed in the right hand side of Fig. 7.3 that the higher level has been reached if $E(T) > E_0 = (1/2)(E_1 + E_2)$. If $E(T) > E_0$ and $\mathcal{P}_2(T) > 1/2$, then the readout $[E]$ points to a transition, and the real behavior of the system $[\mathcal{P}_2]$ is in accord with this information (the information is valid). The probability of this is presented by large points in the right hand side of Fig. 7.3. The middle sized points represent the probability that $E(T) < E_0$ and $\mathcal{P}_2(T) < 1/2$, so the readout points to the absence of the transition, and the real behavior of the system is in accord with this information (i.e., the information is valid). In both diagrams the curves, and correspondingly the probabilities, are approximately equal for $4\pi T_{lr}/T_R = 4/3$. This characterizes the intermediate regime. The probability that the information is false (i.e., the readout points to the transition but the system stays on the first level, or vice versa) is represented by small points in the right hand diagram. This type of error is the measurement noise. This noise increases with fuzziness, achieving 20 %. For the intermediate regime the noise is about 10 %.

7.3.4. LONG CONTINUOUS MEASUREMENT

We have discussed the description of the continuous fuzzy measurement of energy in the non-selective way. This gives interesting results in the case of moderate fuzziness or, in other words, for the duration of the measurement T of the order of the level resolution time T_{lr}. It is difficult to investigate in the same way the case where $T_{\mathrm{R}} \ll T_{\mathrm{lr}} \ll T$. We restrict ourselves to some comments about this regime.

In this case small errors related to the approximate character of the calculation grow exponentially, so this solution is inappropriate for the analysis, and the conclusions made above are not valid. The evolution is qualitatively different in this case. Under these conditions the system 'forgets' because of the measurement its initial conditions.

The mechanism of 'forgetting' is loss of coherence leading to a mixture instead of a pure state. If the duration T of the measurement is not too long the result of the measurement and the evolution of the system may be adequately represented (as has been done above) by a single readout $[E]$ and the resulting pure final state $|\psi_T^{[E]}\rangle$. A single readout is a set of measure zero and, strictly speaking, is characterized by a probability of zero. One has to consider a band of readouts instead, and integrate the probability density $P[E]$ over this band to obtain a non-zero probability. However, close measurement readouts $[E]$ lead to close final states of the system if the measurement is not too long. Therefore the final state corresponding to the band of readouts will be the same pure state $|\psi_T^{[E]}\rangle$ which corresponds to each of the readouts in the band.

This argument is not valid, however, for a long measurement. In this case even very close readouts $[E]$ lead to different final states $|\psi_T^{[E]}\rangle$. Therefore even a very thin band of readouts gives a mixed final state. As T increases, this mixed state becomes universal, not depending on the initial state of the system.

Indirectly, the latter statement is confirmed by the non-selective description of the measurement with the help of the density matrix. It has been shown in Sect. 5.2 that this density matrix satisfies a master equation having the form of the Liouville equation but with an additional damping term proportional to the double commutator $[A[A\rho]]$. In our case this equation reads as follows:

$$\dot{\rho} = -\frac{i}{\hbar}[H_0 + V, \rho] - \frac{\kappa}{2}[H_0[H_0\rho]], \tag{7.25}$$

and can easily be solved for a resonantly driven two-level system. This gives for $8\pi T_{\rm lr} > T_{\rm R}$ (the Rabi regime)

$$\rho_{11} - \rho_{22} = e^{-(t/4T_{\rm lr})}(a \cos \Omega t + b \sin \Omega t). \tag{7.26}$$

Here the matrix elements are taken between the states (7.6),

$$\Omega = \sqrt{(2v)^2 - (2T_{\rm lr})^{-2}}$$

and the parameters a, b are determined by initial conditions. It is evident that the difference $\rho_{11} - \rho_{22}$ owing to the non-symmetrical initial conditions disappears during time of the order of $T_{\rm lr}$.

7.4. Discussion

We have obtained the following results. It is possible to visualize the transition between levels of an externally driven single two-level system by a continuous measurement of energy, if its fuzziness is appropriately chosen. It is best to show this within the phenomenological RPI approach or to use the equivalent Schrödinger equation with a complex Hamiltonian. The information obtained about the transition between levels then has a restricted, but not small, reliability (measurement noise is about 10% for a moderate fuzziness) but may be obtained only at the price of changing the probability of the transition.

Measurements of this type may be realized by a long series of sufficiently weak interactions of the system with a quantum mechanical meter system. This will be considered in Chapter 8.

7.5. Comment

The RPI approach has been applied to the measurement of energy in (OPT93; TPO95). However, an *ad hoc* assumption was made in those works that the measurement readout is a constant curve $E(t) = E_n = \text{const}$. This is too restrictive and misleading for a continuous fuzzy measurement. As a result the analysis given in these papers was correct only in the case of strong (hard) measurements resulting in the quantum Zeno effect. In this case the measurement readout actually (with high probability) is close to one of the levels (see the argument in Sect. 7.2), so the assumption made *a priori* may be justified by the strict analysis. However, the Rabi regime of very weak measurement could not be considered completely and the most interesting intermediate regime could not be even discovered.

Correct and complete consideration of the continuous measurement of energy in a two-level driven system has been performed in a series of works (AM97; AMN97; AM98). The results are sketched in the present chapter.

The majority of authors studying continuous quantum measurements base their investigations on particular micro-models for the measuring device, see (Per87; WM85; Mil88; GM93; Per93) for the measurement of the energy of a two-level system. The results cannot be compared with those reported here because in all these papers only a non-selective description of measurements was applied, so the problem of visualization of the transition could not be posed nor even formulated.

Chapter 8

Monitoring by a series of observations

In this chapter we shall present a concrete model of a continuous fuzzy measurement of an observable in a two-level system. In a special case of minimally disturbing continuous measurement of energy this may be a realization of the process described phenomenologically in Chapter 7. In the general case the process to be considered may realize the non-minimally disturbing continuous measurement (in the sense of Sect. 5.1) of an arbitrary observable in a two-level system.

The measurement will be realized as a series of short weak interactions of the system with a subsidiary system, a meter. The idea of this realization is that each short interaction plays the role of a very fuzzy short measurement. We shall call it an observation to distinguish it from the whole process of continuous measurement (a long chain of observations). Each observation supplies some very vague information about the system state, this information being recorded in the final state of the meter. Therefore the state of the system is disturbed very weakly. Nevertheless, the whole series of observations realizes a continuous fuzzy measurement with a finite fuzziness, which depends on the frequency of 'elementary' observations and the strength of each of them.

We shall consider such a measurement in a selective way so that the behavior of the system will be represented with its dependence on the results of the measurement taking into account. The analysis shows that a series of weak observations gives just the same law of evolution for the measured system which has been described previously by the restricted path integral or complex Hamiltonian. Moreover, a series of weak observations automatically leads to the simplest 'Gaussian' weight functional or quadratic imaginary term in the Hamiltonian. This is an analogue of the Central Limiting Theorem, but in the quantum context. It is quite probable that the same may be proved in a wider scope of situations. If this is the case, the Gaussian law of evolution of a continuously measured system (described in a general form in Chapter 5.1) may be considered universal just as the Gaussian (normal) distribution is universal in probability theory.

The plan of the chapter is the following. First (Sect. 8.1) we show how the continuous measurement is arranged by a long series of observations. Then (Sect. 8.2) we consider a simple model of a single observation and find out how the system state changes in the course of this observation. In Sect. 8.3 we derive the law of evolution of the system under a series of observations, and in (Sect. 8.4) we demonstrate that this law coincides with the phenomenological description of a continuously measured system.

8.1. The general scheme

Consider a quantum system with two energy levels E_1 and E_2 which is under the influence of a driving field. The total Hamiltonian is $H = H_0 + V$, where

$$H_0 = E_1|1\rangle\langle 1| + E_2|2\rangle\langle 2|.$$

The continuous measurement in this two-level system is assumed to be constituted of many short interactions of duration $\delta\tau$ of the two-level system with another quantum system called the *meter*. Such a short interaction plays the role of observation of our system with information recorded in the state of the meter. Single interactions (observations) are separated by periods, of duration τ, of evolution governed by the Hamiltonian H. After each interaction the meter is in one of two states, and this final state carries some information about the measured system. When the next observation starts, the meter is again in the initial state. This may mean that the meter is restored to the initial state (after registering its final state) or that another meter identical to the first one and in the same initial state performs the next observation.

Thus each observation gives one of two possible results which will be denoted by '+' or '−'. This scheme of arrangement of the continuous measurement is a generalization of one proposed in (AMN97) where s-wave scattering of an electron (taken as a meter) by an atom was considered with two possible results: the electron may be deflected (positive result) or not (negative result).

A long series of such interactions will be shown to lead to continuous measurement of some observable of the two-level system. First we shall see that this may be interpreted as a measurement of an observable p connected in a simple way with the probability to obtain positive result of the observation. Namely, the quantum mechanical mean value $\langle p\rangle$ of the operator p in the given state is equal to the probability of obtaining a positive result if the observation is performed in this state. Later we shall see that the same process may be interpreted as a continuous measurement of any other observable A, provided that it has the same eigenvectors as the operator p.

The evolution of the two-level system in a series of observations will be calculated and compared with the evolution predicted by the phenomenological restricted-path-integral (RPI) approach. The evolution depends on the result of a series of observations which may be represented as a series of pluses and minuses (positive and negative results of the observations). The crucial role is played by the number of pluses compared with the complete number of observations.

In order to interpret a long series of observations as a realization of a continuous measurement we shall divide it into shorter (but nevertheless long) series, each series containing N observations and having a duration Δt. Such an N-series will be used to elaborate a single number $n(t) = N_+/N$, where N_+ is the number of positive results in the N-series of observations. The number $n(t)$ will be associated with the instant t when the given N-series began and identified with the result of the measurement of the operator p at the given time. Doing the same for each of N-series forming the complete series of observations, we have numbers

$$n(0), n(\Delta t), n(2\Delta t), \ldots$$

which will be interpreted as a representation of a curve $[n] = \{n(t)|0 \leq t \leq T\}$, the result of continuous measurement of the operator p.

The main point of the whole consideration is the following statement: evolution of the two-level system in a long series of observations is in agreement with the phenomenological description of continuous measurement of the operator p leading to the given measurement readout $[n]$. The same evolution may be interpreted as continuous measurement of another operator A leading to the readout $[a]$. This re-interpretation is possible if A has the same eigenvectors as p. In this case the number $a(t)$ is expressed in a specified way in terms of $n(t)$.

8.2. Model of the observation

Our aim is now to discuss how the observation mentioned in Sect. 8.1 may be realized by an interaction of the two-level system with a subsidiary system, the meter.

Let our system with Hamiltonian $H = H_0 + V$ interact with the meter, which has Hamiltonian H_M, via the interaction Hamiltonian V_I. Then the Schrödinger equation for the *composite quantum system* (the 2-level system plus the meter) with state $|\Psi\rangle$ in the product space is

$$i\hbar|\dot{\Psi}\rangle = (H + H_M + V_I)|\Psi\rangle. \tag{8.1}$$

We turn to a short period of interaction leading to an observation. The system and the meter are initially uncorrelated, hence the state $|\Psi(0)\rangle$ of the combined system is a product of the system state $|\psi\rangle$ and the normalized meter state $|\Phi_0\rangle$,

$$|\Psi(0)\rangle = |\psi\rangle|\Phi_0\rangle. \tag{8.2}$$

Restricting ourselves to a time-independent interaction operator V_I and assuming that the duration $\delta\tau$ of the interaction is short so that the dynamics owing to $H + H_M$ is negligible during this period, we have for the state at the end of the interaction, as consequence of the unitary development,

$$|\Psi(\delta\tau)\rangle = e^{-\frac{i}{\hbar}\delta\tau V_I}|\Psi(0)\rangle \approx \left(1 - \frac{i}{\hbar}V_I\delta\tau\right)|\Psi(0)\rangle. \tag{8.3}$$

The states of the meter corresponding to a positive or negative result of the observation are denoted by $|\Phi_+\rangle$ and $|\Phi_-\rangle$ respectively (Fig. 8.1). They

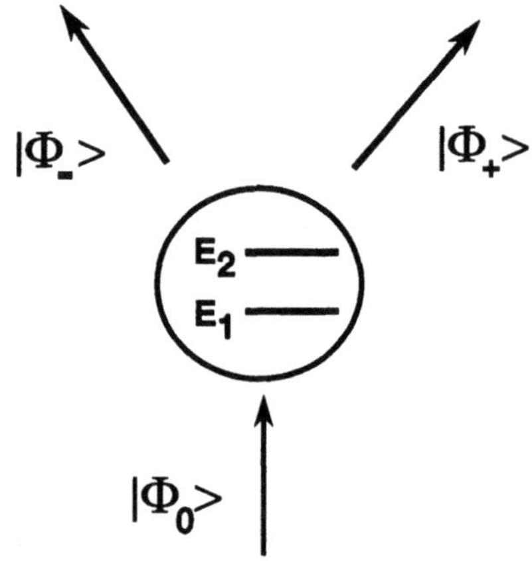

Figure 8.1. Realization of an observation. A two-level system interacts with a subsidiary system, the meter. Before the observation the states $|\Phi_0\rangle$ of the meter and $|\psi\rangle$ of the system are uncorrelated. They become entangled during the interaction: If the state $|\Phi_+\rangle$ of the meter is observed (a positive result of the observation), the system is transferred to the state $|\psi_+\rangle$. Accordingly, the state $|\Phi_-\rangle$ of the meter after the observation is correlated with the state $|\psi_-\rangle$ of the system.

form an orthogonal basis in the meter space. Then, multiplying Eq. (8.3) by the decomposition of unity,

$$\mathbf{1} = |\Phi_+\rangle\langle\Phi_+| + |\Phi_-\rangle\langle\Phi_-|,$$

we have

$$|\Psi(\delta\tau)\rangle = |\Phi_+\rangle|\psi_+\rangle + |\Phi_-\rangle|\psi_-\rangle, \tag{8.4}$$

where

$$\begin{aligned} |\psi_+\rangle &= \langle\Phi_+|\left(1 - \frac{i}{\hbar}\delta\tau V_I\right)|\Phi_0\rangle|\psi\rangle, \\ |\psi_-\rangle &= \langle\Phi_-|\left(1 - \frac{i}{\hbar}\delta\tau V_I\right)|\Phi_0\rangle|\psi\rangle. \end{aligned} \tag{8.5}$$

The interaction converts the composite system into the entangled state (8.4). Measuring the subsystem meter by a macroscopic measuring apparatus finds it in one of two states $|\Phi_+\rangle$ or $|\Phi_-\rangle$. The 2-level system then turns out to be in the corresponding state $|\psi_+\rangle$ or $|\psi_-\rangle$.

This is nothing else than decoherence of the two-level system caused by its entanglement with the meter. The essential features of decoherence were discussed (although in a simpler situation) in Sects. 2.1, 2.2.1. According to the arguments of Sect. 2.2.1, decoherence is the result of the entanglement and does not depend, in fact, on whether the second component of the entanglement (the meter in our case) is actually measured or not. We may describe decoherence in a non-selective way, using the density matrix and taking into account both results of the measurement ('+' and '−') simultaneously, or describe it selectively, considering each of these results separately. These are different pictures of the same physical phenomenon. We now prefer the selective description.

As a consequence of Eq. (8.5) we have

$$|\psi_+\rangle = u_+|\psi\rangle, \quad |\psi_-\rangle = u_-|\psi\rangle,$$

where

$$\begin{aligned} u_k &= \langle\Phi_k|e^{-\frac{i}{\hbar}\delta\tau V_I}|\Phi_0\rangle \\ &\approx \langle\Phi_k|\left(1 - \frac{i}{\hbar}\delta\tau V_I\right)|\Phi_0\rangle \end{aligned}$$

($k = \pm$) are operators in the state space of the two-level system. They may be realized, for example, as two-dimensional matrices acting in the energy representation of the system. They can then be expressed through the Pauli matrices and the unit 2×2 matrix. The norms of the states $|\psi_\pm\rangle$ represent probabilities of the corresponding results of the observation according to

$$\text{Prob}(+) = \frac{||\psi_+||^2}{||\psi||^2}, \quad \text{Prob}(-) = \frac{||\psi_-||^2}{||\psi||^2},$$

so that

$$\text{Prob}(+) = \langle u_+^\dagger u_+\rangle, \quad \text{Prob}(-) = \langle u_-^\dagger u_-\rangle,$$

where the mean value in the state $|\psi\rangle$ is meant.

To express this in a more explicit form let us represent the operators u_k in terms of their 'moduli' and 'phases',

$$u_k = e^{is_k}\sqrt{p_k}, \tag{8.6}$$

where s_k are Hermitian and p_k are positive operators. Then the operators

$$p_k = u_k^\dagger u_k$$

satisfy the relation

$$p_+ + p_- = \mathbf{1},$$

and their mean values represent the corresponding probabilities Prob(±). We shall denote

$$p = p_+.$$

Then the probability of the positive result from the observation is

$$\mathrm{Prob}(+) = \langle p \rangle.$$

It is important for the further calculation to commute the operators u_k and their constituents p_k, s_k defined according to Eq. (8.6). To make the question of commutators clear, let us represent the operators u_k as

$$u_k = c_k + g(a_k + ib_k),$$

where c_k is a number (it may be made real by omitting the inessential phase factor in u_k), a_k and b_k are Hermitian two-dimensional matrices, and $g \ll 1$ represents the smallness of the interaction Hamiltonian V_I. Then we have, up to the first order in g,

$$\sqrt{p_k} = c_k + ga_k, \quad s_k = g\frac{b_k}{c_k},$$

and the commutators of s_k and p_l are of the second order. In addition, the operators $p_+ = p$ and $p_- = 1 - p$ commute with each other precisely. Therefore the factors in the product of N amplitudes u_k (with different k) may be arbitrarily ordered, provided that N is less than or of the order of g^{-1}. If we use the form (8.6) for each of the factors, then the 'phase factors' e^{is_k} and moduli $\sqrt{p_k}$ can also be ordered arbitrarily. We shall use this in our calculation.

8.3. Evolution of the state in a series of observations

The amplitudes u_k show how the state of the system is changed in a single observation depending on the result of the observation $k = \pm$. We shall now use these amplitudes to derive the evolution of the system in a long series of such interactions (observations). Although the observations are separated by periods of free evolution, we shall consider first a series of N observations as if there were no evolution between them. Later we shall add the influence of the free evolution to the amplitude describing the observations.

If several observations follow one another, the influence of the whole series of them is represented by the product of the amplitudes u_k

(with the corresponding k). Therefore if the positive result occurs N_+ times in a N-series of observations the resulting amplitude is equal to

$$\begin{aligned} U_{\text{select}}(N_+, N) &= u_+^{N_+} u_-{}^{N-N_+} \\ &= p^{N_+/2}(1-p)^{(N-N_+)/2} \exp\{iN_+ s_+ + i(N-N_+)s_-\}. \end{aligned} \tag{8.7}$$

Here we have taken into account the approximate commutativity of the operators (see the discussion at the end of Sect. 8.2). The expression (8.7) gives an amplitude for a certain measurement readout, i.e. for a fixed ordered series of pluses and minuses. This is indicated by the index 'select'. It is not difficult, however, to go over to the formula in which all possible orders of pluses and minuses with the fixed total number N_+ of pluses are accounted.

For this goal, let us rewrite the evolution owed to the amplitude (8.7) in a somewhat symbolical form as

$$|\psi\rangle \to U_{\text{select}}(\text{order})|\psi\rangle = U_{\text{select}}|\psi\rangle. \tag{8.8}$$

It is important (and indicated in this formula) that the amplitude (8.7) does not in fact depend on the order of pluses and minuses in the series.

If the order of the positive and negative results is not known or not taken into account we have to go over to a partly non-selective description, summing up over all alternatives in terms of a density matrix,

$$\begin{aligned} |\psi\rangle\langle\psi| &\to \sum_{\text{order}} U_{\text{select}}(\text{order})|\psi\rangle\langle\psi|U^\dagger_{\text{select}}(\text{order}) \\ &= \binom{N}{N_+} U_{\text{select}}|\psi\rangle\langle\psi|U^\dagger_{\text{select}}. \end{aligned} \tag{8.9}$$

The summation here has been reduced to multiplication by the number of different orders because the amplitude (8.7) does not depend of the order. Going back to the evolution of the state we obtain

$$|\psi\rangle \to \sqrt{\binom{N}{N_+}}\, U_{\text{select}}|\psi\rangle. \tag{8.10}$$

Therefore the amplitude describing simultaneously all orders of pluses and minuses with the given N_+ is

$$U(N_+, N) = \sqrt{\binom{N}{N_+} p^{N_+}(1-p)^{N-N_+}}\, \exp\{iN_+ s_+ + i(N-N_+)s_-\}. \tag{8.11}$$

The probability $P(N_+, N)$ of obtaining N_+ positive results in N observations (independently of the order of these results) may be found as the trace of the r.h.s. of Eq. (8.9) if the vector $|\psi\rangle$ is normalized. This amounts to the mean value of the square modulus of the amplitude (8.11):

$$P(N_+, N) = \langle U^\dagger(N_+, N) U(N_+, N)\rangle = \langle \binom{N}{N_+} p^{N_+}(1-p)^{N-N_+}\rangle. \quad (8.12)$$

where the mean value is calculated in the initial state $|\psi\rangle$.

8.4. Comparison with the phenomenological approach

The expression in the r.h.s. of Eq. (8.12) is a known binomial distribution, but for an operator p instead of a number (the probability of an elementary event). It is known that for a large N the binomial distribution may be very well approximated by the normal (Gaussian) distribution. The same also holds for the positive operator p (this may be proved with the help of the spectral decomposition of p). Therefore,

$$\binom{N}{N_+} p^{N_+}(1-p)^{N-N_+} = \mathcal{N}^2 \exp\left[-\frac{N}{2p(1-p)}(n-p)^2\right]. \quad (8.13)$$

Substituting this expression in Eq. (8.11) we have the following expression for our amplitude:

$$U(N_+, N) = \mathcal{N}^2 \exp\left[-\frac{N}{4p(1-p)}(n-p)^2 + iN(n(s_+ - s_-) + s_-)\right]$$

where $n = N_+/N$. Taking into account that the number of observations N is proportional to the duration Δt of the N-series, $\Delta t = N\tau$ and taking into account the free evolution between the observations, we have, after appropriate normalization,

$$U(\Delta t) = \exp\left\{\Delta t\left[-\frac{i}{\hbar}H - \frac{(p-n)^2}{4\tau p(1-p)} + \frac{i}{\tau}(n(s_+ - s_-) + s_-)\right]\right\}.$$

This formula is obtained in the approximation when the commutator of H with the amplitudes u_k may be neglected. This commutator is small because the eigenvalues of the operator p are close, owing to the weakness of the observation. A more detailed consideration shows that the commutator may be neglected if the N-series is not too long (see Sect. 8.5 for details).

To obtain the amplitude corresponding to a long period we have to multiply, in the appropriate order, many amplitudes of this form for subsequent time periods of duration Δt (which is assumed small). As a result we shall have an ordered exponential of a time integral. We shall represent it in the form

$$U_T = \mathcal{T} \exp\left\{\int_0^T dt \left[-\frac{i}{\hbar}H - \sigma(t)\left(\frac{p - n(t)}{\Delta p}\right)^2 - \frac{i}{\hbar}\left(\frac{n(t) - p_0}{\Delta p}B + C\right)\right]\right\}, \tag{8.14}$$

where $\mathcal{T}$ means time ordering of the exponent, and where we have introduced the following notations:

$$\sigma = \frac{\Delta p^2}{4\tau p(1-p)}, \quad B = -\frac{\hbar \Delta p}{\tau}(s_+ - s_-), \quad C = -\frac{\hbar}{\tau}[p_0\, s_+ + (1-p_0)s_-].$$

We have used here, for further convenience, the parameters

$$\Delta p = p_2 - p_1, \quad p_0 = \frac{1}{2}(p_1 + p_2) \tag{8.15}$$

composed of the eigenvalues p_1, p_2 of the operator p.

The integral (8.14) represents the evolution which has been equivalently described in Sect. 5.1 by the restricted path integral with a weight functional of the type of Eq. (5.19). In our case the measured operator is p and measurement readout $[n]$. According to the definition of Sect. 5.1, the measurement is non-minimally disturbing, the redundant disturbance being represented by the operators B and C. Note that the coefficient σ determining fuzziness of the measurement is in this case an operator (a function of the operator p) rather than a number. However it may be shown that the operator σ may be replaced by a number if $Np_0 \gg 1$. This condition must be fulfilled in any real situation because it is necessary for a typical N-series to contain at least one positive result.

Let us analyze the operator p which is measured in our case. If it has eigenvalues p_1, p_2 and the corresponding eigenvectors $|\psi_1\rangle$ and $|\psi_2\rangle$ (all of them may be calculated on the basis of the preceding definitions), then

$$p = p_1 P_1 + p_2 P_2,$$

where

$$P_1 = |\psi_1\rangle\langle\psi_1|, \quad P_2 = |\psi_2\rangle\langle\psi_2|.$$

The interpretation of these is following. If the system is in the state $|\psi_1\rangle$ (correspondingly $|\psi_2\rangle$), the probability that the observation will give positive result is equal to p_1 (correspondingly p_2). A simple calculation shows then that the statistical average of different values of $n = N_+/N$ obtained in an N-series starting from the state $|\psi\rangle$ is equal to the quantum mechanical mean value of the operator p. Indeed, with Eq. (8.12) we have

$$\overline{n} = \sum_{N_+=0}^{N} \frac{N_+}{N} P(N_+, N) = \langle p \rangle. \tag{8.16}$$

This gives one more justification that the process really does lead to continuous measurement of the operator p, with $n(t)$ being the continuous measurement's readout. However, the same process may, in fact, be interpreted as measurement of another operator, provided that it has the same eigenvectors as the operator p. Let

$$A = a_1 P_1 + a_2 P_2.$$

Then the operators A and p and their mean values are connected by the relations

$$\frac{A - a_0}{\Delta a} = \frac{p - p_0}{\Delta p}, \qquad \frac{\langle A \rangle - a_0}{\Delta a} = \frac{\langle p \rangle - p_0}{\Delta p},$$

where

$$\Delta a = a_2 - a_1, \qquad a_0 = \frac{1}{2}(a_1 + a_2).$$

This means that the result $n(t)$ of the measurement of the operator p must be identified with the result $a(t)$ of the measurement of the operator A, provided that the number $a(t)$ is expressed in terms of $n(t)$ by an analogous formula:

$$\frac{a(t) - a_0}{\Delta a} = \frac{n(t) - p_0}{\Delta p}.$$

Substituting the expressions following from these formulas for p and $n(t)$ in Eq. (8.14) we may represent the same law of evolution in the form

$$U_T = T \exp\left\{ \int_0^T dt \left[-\frac{i}{\hbar} H - \sigma(t) \left(\frac{A - a(t)}{\Delta a} \right)^2 - \frac{i}{\hbar} \left(\frac{a(t) - a_0}{\Delta a} B + C \right) \right] \right\}. \tag{8.17}$$

Comparing this formula with the general results at the end of Sect. 5.1, we see that the process considered may be interpreted as a non-minimally disturbing continuous measurement of the operator A. In particular, if $|\psi_1\rangle$, $|\psi_2\rangle$ are the energy eigenvectors the process may be interpreted as a non-minimally disturbing continuous measurement of energy.

8.5. Parameters of the model

Fuzziness of the continuous measurement (8.14) depends on the parameter σ. As usual, it is convenient to use instead of it a parameter having the dimension of time, the level resolution time (cp. Sect. 7.2.2)

$$T_{\rm lr} = \frac{1}{\sigma} = \tau \frac{4p_0(1-p_0)}{\Delta p^2}. \tag{8.18}$$

It is seen from Eq. (8.18) that a large level resolution time $T_{\rm lr}$ (and therefore a fuzzy continuous measurement) is provided by small Δp and large τ. This means that each elementary observation must be fuzzy and the repetition of observations not too frequent. The series of elementary observations may be considered as a continuous measurement if the period τ between single observations is much less than the characteristic time scale of the continuous measurement $T_{\rm lr}$. We see from Eq. (8.18) that this occurs if

$$\frac{p_0}{\Delta p^2} \gg 1 \tag{8.19}$$

(we have assumed that $1 - p_0$ is of the order of unity).

The commutator (neglected in the above consideration) between H and the amplitudes u_k is actually small, because the eigenvalues of the operator p are close, owing to the weakness of the observation. A more detailed consideration shows that the commutator may be neglected if the N-series is not too long, thus

$$N^2 \ll \frac{\sqrt{p_0}}{\Delta p v \tau}, \tag{8.20}$$

where $\hbar v$ is the maximum absolute value of matrix elements of the interaction Hamiltonian V_I. The parameter $v\tau = \pi\tau/T$ is assumed to be small. The ratio $\Delta p^2/p_0$ is also small if τ is less than the level resolution time $T_{\rm lr}$ (cp. Eq. (8.19)). Therefore the number N may be large, even despite the restriction (8.20).

A more concrete model of the observation leading to a minimally disturbing measurement of energy has been considered in (AM98). In this model it has been assumed that in the case of the negative result of the observation the final state of the meter coincides with its initial state, $|\Phi_-\rangle = |\Phi_0\rangle$. This is fulfilled for the example in the experimental setup based on scattering (AMN97), where the outgoing state of an electron for the negative result of the observation (no deflection) coincides with its ingoing state (before scattering). The consequence is the orthogonality of $|\Phi_0\rangle$ and $|\Phi_+\rangle$, therefore we have, up to a phase factor $(-i)$,

$$\begin{aligned} |\psi_+\rangle &= \frac{\delta\tau}{\hbar}\langle\Phi_+|V_I|\Phi_0\rangle|\psi_0\rangle, \\ |\psi_-\rangle &= \left(1 - \frac{i}{\hbar}\delta\tau\langle\Phi_0|V_I|\Phi_0\rangle\right)|\psi_0\rangle. \end{aligned} \tag{8.21}$$

The interaction Hamiltonian V_I was taken in (AM98) in the factorized form, with the factor referring to the measured system being diagonal in the energy representation:

$$V_I = V_{I0}(1 - \lambda\sigma_3). \tag{8.22}$$

Here V_{I0} acts only on the state of the meter, σ_3 is the Pauli matrix acting in the two-dimensional space of energy levels 1 and 2, and λ is a number much smaller than 1. Then the final states of the measured system after the short interaction with the meter are

$$\begin{aligned} |\psi_+\rangle &= a(1 - \lambda\sigma_3)|\psi_0\rangle, \\ |\psi_-\rangle &= e^{ib\lambda\sigma_3}|\psi_0\rangle, \end{aligned} \tag{8.23}$$

where

$$a = \frac{\delta\tau}{\hbar}\langle\Phi_+|V_{I0}|\Phi_0\rangle, \quad b = \frac{\delta\tau}{\hbar}\langle\Phi_0|V_{I0}|\Phi_0\rangle. \tag{8.24}$$

The amplitudes describing the observation are then

$$u_+ = a(1-g)|1\rangle\langle 1| + a(1+g)|2\rangle\langle 2|, \quad u_- = e^{ibg}|1\rangle\langle 1| + e^{-ibg}|2\rangle\langle 2|. \tag{8.25}$$

Making use of Eq. (8.25), we have from (8.15)

$$p_0 = |a|^2(1+\lambda^2), \quad \Delta p = 4\lambda|a|^2. \tag{8.26}$$

The probabilities p_1 and p_2 are therefore close if λ is chosen small enough.

The level resolution time of (8.18) becomes

$$T_{\mathrm{lr}} = \frac{\tau}{4\lambda^2|a|^2}. \tag{8.27}$$

Thus $1/4\lambda^2|a|^2$ is the critical number of observations leading to the resolution of energy levels (see a detailed discussion of this resolution in Sect. 7.2). The phase factor in Eq. (8.23) is negligible for a continuous measurement of duration T if

$$\lambda b \ll \frac{\tau}{T}. \tag{8.28}$$

This is valid for T of the order of T_{lr} if $b \ll 4\lambda|a|^2$. This condition is satisfied for the system considered in (AMN97).

8.6. Discussion

We have obtained the following results:

i) Continuous fuzzy measurement of an arbitrary observable of a two-level system may be realized by a long series of sufficiently weak interactions of the system with a quantum mechanical meter system, which is then measured by a macroscopic apparatus. The discussion of the dynamics shows that the model reproduces phenomenologically obtained results.
ii) The analysis leads to an operational interpretation of the readout $a(t)$ for the measured operator A. It turns out to be the estimate of the mean value $\langle A \rangle$ of the observable A with respect to the system state $|\psi(t)\rangle$ at that time. The precision of the estimate depends on the fuzziness of the measurement.

The realization of a continuous measurement by a long series of elementary observations generalizes the experimental setups for measurement of energy proposed in (BHL$^+$90; AMN97). However, the aim of the paper (BHL$^+$90) (see also (SBS92) for more detailed discussion of the same setup) was not the investigation of a continuous measurement, but the decoherence of levels resulting if the measurement is long enough (in our notations, longer than the decoherence time, $T \gg T_{\text{lr}}$).

It is interesting that for a wide class of elementary observations their repetition leads to the simplest form of the continuous measurement characterized by a Gaussian functional in the restricted path integral or by a quadratic imaginary term in the Hamiltonian. The resulting measurement is characterized in the minimally disturbing case by only one single parameter, namely the fuzziness. In the non-minimally disturbing case two other parameters (operators) arise corresponding to real additional terms in the Hamiltonian, but the weight functional remains Gaussian.

The interaction of the measured system with the meter may depend on many parameters, but only few combinations of these parameters, namely fuzziness and two operators responsible for disturbing it, arise in the description of the resulting continuous measurement. This is quite analogous to the the well known result in probability theory (the famous central limiting theorem) that the result of a long series of random events is normally distributed independently of the distributions of single events in the series. Therefore, the considerations here formulate and prove the simplest *quantum analog of the classical central limiting theorem.*

The same must be valid also in a wider range of situations. For example, the instants when observations are performed, may be not prescribed precisely, but chosen randomly. Moreover, the measured system may be more complicated than a two-level system.

It is more difficult to consider a continuous measurement realized as a series of different short interactions (observation) rather than as a repetition of the same interaction. The amplitudes representing these observations are not obliged then to commute with each other, therefore the consideration is more complicated. It may be expected, nevertheless, that the result may be represented analogously to Eq. (8.17) but with the sum of the terms in the exponent representing different measured observables. This would correspond to a simultaneous measurement of several observables, as has been commented at the end of Sect. 5.1.3, see Eqs. (5.22, 5.23). The definition of the path integral would, in this case, be more complicated; however, it may be expected that a quantum generalization of the Central Limiting Theorem may be formulated and proven in this way.

Chapter 9

The Quantum Zeno effect in a passive setup

In the present chapter we shall give one more, but a somewhat unusual example of a continuous fuzzy measurement of energy (see (AMP99)).[1] Its peculiarity is that it is a measurement with a negative result (null measurement). The measuring device in a null measurement is constructed so as to discover some object or event, but the measurement gives a negative result. Nevertheless, the null measurement influences the system. An example to be considered is peculiar in that the negative-result measurement is continuous.

We consider a driven 2-level system with one level demonstrating spontaneous decay to an otherwise uncoupled third level. Rabi transitions to the unstable level, which might be induced by the driving field, turn out to be strongly damped because of instability of the target level. We shall argue that this is a consequence of the quantum Zeno effect, the very possibility of the decay of the second level being a sort of continuous measurement.

The given setup will be analyzed by the usual quantum mechanical methods. However, to make the nature of the phenomenon clearer, a comparison of this setup with the known scheme of repeated projection measurements is given. Furthemore, it is shown that the measurement of this type may be included in the general scheme of the phenomenological restricted-path-integral theory of continuous measurements. The considered system confirms an important role of null measurements and may serve as a good example for a truly continuous measurement (as compared to a repeated one).

9.1. Active and passive setups for the Zeno effect

The quantum Zeno effect (see Sect. 4.1) is one of the specific quantum effects which have gained much interest, in particular in connection with

[1]This chapter is reprinted, with appropriate changes, from "Physics Letters A", Volume 261, J.Audretsch, M.Mensky and A.Panov, Zeno effect preventing Rabi transitions onto an unstable energy level, Pages 44–50, Copyright 1999, with the permission of Elsevier Science.

the quantum theory of measurements. It goes back to successive or truly continuous measurements of the system and depends on the strength of the measurement. Generally speaking, it consists in slowing down induced quantum transitions as a result of repeated measurements of the system. In the limit of a hard continuous measurement, transitions of the system are inhibited, and the system is frozen on a level.

It is characteristic for the Zeno effect is that it is caused by measurement. A prevention of a transition must not necessarily be caused by a measurement. It may also happen that the Zeno effect (induced by a measurement) is accompanied by other influences. In the following we discuss a new demonstration of a pure Zeno effect.

A well known particular example of the Zeno effect (see Sect. 4.1) is slowing down of Rabi transitions between the two energy levels $|1\rangle$ and $|2\rangle$ of a 2-level system, which are induced by a resonant driving laser field, as the result of frequently repeated projective energy measurements. In the well known experiment of Itano *et al.* (Sect. 4.1.2) the repeated energy measurement is realized for ions with the help of periodic short pulses of another laser field inducing optical transitions from level $|1\rangle$ to a third level $|3\rangle$ with a subsequent spontaneous decay back to the initial level $|1\rangle$ (the V-configuration). The photon emitted by spontaneous emission is monitored. If the ion is at level $|1\rangle$ it will emit photons. This does not happen if it is at level $|2\rangle$. In this sense the fluorescence photons contain information about the state of the ion. The $|1\rangle \rightarrow |3\rangle$ transition together with the emission of the fluorescence photon represents a non-destructive projection measurement of level $|1\rangle$. As a result the transition $|1\rangle \rightarrow |2\rangle$ is hindered. This freezing at a level is a pure demonstration of the quantum Zeno effect caused by repeated almost instantaneous projection measurements (cp. (BHS96), where also a more complete list of the literature can be found).

The aim of this chapter is to show that the Zeno effect may be demonstrated in a much simpler way for a driven 2-level system with one level showing spontaneous decay to an otherwise uncoupled third level. At the same time this system may serve as an example for a truly continuous quantum measurement. In addition, it visualizes the role of null measurements.

We consider the simple quantum system presented in Fig. 9.1a. It is a 2-level system, with levels $|1\rangle$ and $|2\rangle$, which is subject to the influence of a resonant driving field V generating Rabi oscillations between $|1\rangle$ and $|2\rangle$. Initially the system is at level $|1\rangle$, and the driving field induces a transition to level $|2\rangle$. However, level $|2\rangle$ is assumed to be unstable. It decays rapidly by spontaneous decay, with a relaxation rate γ, to a third level $|3\rangle$ with the emission of a photon. Accordingly, if the system reaches level $|2\rangle$ it very quickly transits further to level $|3\rangle$. Therefore, if no photon is emitted this failure is a sign that the system stays perpetually at level $|1\rangle$. Consequently we may consider this setup as a realization of a continuous measurement of the energy level $|1\rangle$ in a passive way leading to null results (in contrast to the non-continuous and active scheme realized by Itano *et al.*).

What type of dynamical behavior is to be expected? Naïvely one could imagine that the system is transferred with the first Rabi pulse to level $|2\rangle$ from where it quickly decays to level $|3\rangle$. The opposite turns out to be the case. The transition $|1\rangle \rightarrow |2\rangle$ is slowed down and the effectiveness for this freezing at the initial level $|1\rangle$ actually increases with the effectiveness of the spontaneous decay of level $|2\rangle$, i.e., with an increasing relaxation rate γ. This may seem to be counter-intuitive or paradoxical. But in fact it is no surprise, because in a situation in which a continuous quantum measurement is realized as a result of the coupling to the environment, the quantum Zeno effect is to be expected. We will show that this behavior of the system can, indeed, very easily be understood as another experimentally accessible example of a pure manifestation of the Zeno effect. That this effect is, in our case, based only (i) on null-measurements and (ii) on genuinely continuous measurements, makes it even more interesting from the conceptual point of view.

We start in Sect. 9.2 with a quantum optical description of the microphysical dynamics underlying this Zeno effect. We compare it in Sect. 9.3 with the phenomenological approach based on repeated measurements. In Sect. 9.4 we give a treatment within a phenomenological theory of continuous quantum measurements.

9.2. Quantum optical treatment

We are dealing with a 3-level system which is under the influence of a driving field and in interaction with the electromagnetic vacuum, causing the spontaneous emission of a photon. An adequate treatment of the complete underlying Schrödinger dynamics of the open system is rather involved. Let us therefore introduce, already at the level of the microphysical quantum optical treatment, some phenomenological elements.Instead of the 3-level system we study an equivalent 2-level system consisting only of the levels $|1\rangle$ and $|2\rangle$ with energies $\hbar\omega_1$ and $\hbar\omega_2$, and represent the instability of level $|2\rangle$ by an imaginary term in the energy of this level (Fig. 9.1 b).

We write the Hamiltonian of the system as $H = H_\gamma + V$ with

$$H_\gamma|1\rangle = \hbar\omega_1|1\rangle, \quad H_\gamma|2\rangle = \hbar(\omega_2 - i\gamma)|2\rangle,$$
$$\langle 1|V|2\rangle = \langle 2|V|1\rangle^* = \hbar\Omega\, e^{i(\omega_2-\omega_1)t}. \tag{9.1}$$

Let us introduce amplitudes $a_1(t)$, $a_2(t)$ as follows:

$$|\psi\rangle = a_1(t)\, e^{-i\omega_1 t}|1\rangle + a_2(t)\, e^{-i\omega_2 t}|2\rangle. \tag{9.2}$$

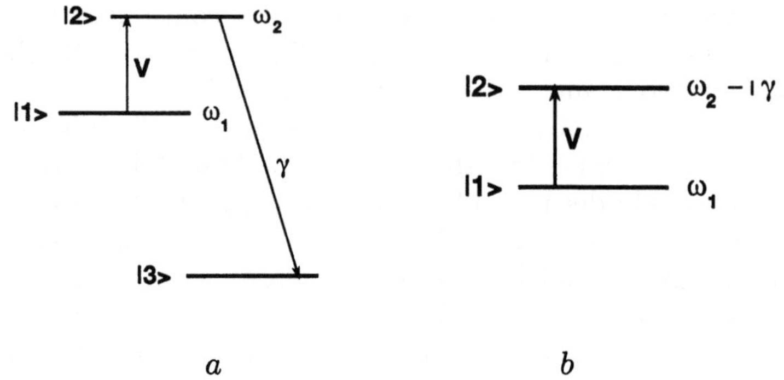

Figure 9.1. A 3-level system with an initial state $|1\rangle$ and a Rabi transition to state $|2\rangle$, from where it may decay only to level $|3\rangle$ accompanied by the emission of a photon (Fig. a). The equivalent 2-level system has a complex energy of the unstable level $|2\rangle$ (Fig. b).

The dynamics

$$|\dot{\psi}\rangle = -\frac{i}{\hbar}(H_\gamma + V)|\psi\rangle \tag{9.3}$$

then results in the following differential equations for the amplitudes

$$\dot{a}_1 = -i\Omega a_2, \quad \dot{a}_2 = -i\Omega a_1 - \gamma a_2 \,. \tag{9.4}$$

The solution is

$$\begin{aligned} a_1 &= e^{-\frac{\gamma}{2}t}\left[a_1(0)\cosh\Omega_\gamma t + \frac{1}{\Omega_\gamma}\left(\frac{\gamma}{2}a_1(0) - i\Omega a_2(0)\right)\sinh\Omega_\gamma t\right], \\ a_2 &= e^{-\frac{\gamma}{2}t}\left[a_2(0)\cosh\Omega_\gamma t - \frac{1}{\Omega_\gamma}\left(i\Omega a_1(0) + \frac{\gamma}{2}a_2(0)\right)\sinh\Omega_\gamma t\right], \end{aligned} \tag{9.5}$$

where we have introduced

$$\Omega_\gamma = \sqrt{(\gamma^2/4) - \Omega^2}\,. \tag{9.6}$$

Under the condition that the system is initially ($t = 0$) on level $|1\rangle$, we find

$$\begin{aligned} a_1(t) &= e^{-\frac{\gamma}{2}t}\left(\cosh\Omega_\gamma t + \frac{\gamma}{2\Omega_\gamma}\sinh\Omega_\gamma t\right), \\ a_2(t) &= e^{-\frac{\gamma}{2}t}\left(-i\frac{\Omega}{\Omega_\gamma}\sinh\Omega_\gamma t\right). \end{aligned} \tag{9.7}$$

Depending on the relative strength of the two parameters γ and Ω one can discriminate different regimes of behavior. If the influence of the driving field is strong in comparison with the strength of the spontaneous decay, i.e., for $\Omega \gg \gamma$, we are in the Rabi regime, where the behavior of the system shows the usual Rabi oscillations modified by some damping:

$$a_1(t) = e^{-\frac{\gamma}{2}t} \cos \Omega t, \quad a_2(t) = -ie^{-\frac{\gamma}{2}t} \sin \Omega t. \tag{9.8}$$

We see that the system, being initially at level 1, will be at level 2 after the Rabi period $T_R = \pi/2\Omega$.

For our purpose it is the opposite regime with strong coupling, $\gamma \gg \Omega$, to the environment which is of interest. In this case the probability of the permanent survival, $P_{ps}(T)$, of the system at level $|1\rangle$ during a sufficiently long time T turns out to be:

$$P_{ps}(T) = |a_1(T)|^2 = e^{-2\Omega^2 T/\gamma}. \tag{9.9}$$

Because $\gamma \gg \Omega$ we find damping of the Rabi oscillations. In the limit $\Omega/\gamma \to 0$ we have $P_{\rm ps}(T) \to 1$ for any fixed T, and the system is frozen at level $|1\rangle$. This is a manifestation of the Zeno effect.

We recall that the Zeno effect is the phenomenon that continuous measurement or many consecutive quantum measurements lead to damping and, in the limit, to the freezing of the evolution of the measured system. In our case we have a continuous measurement because of the coupling to the environment via spontaneous decay. And we have just shown that it results in the freezing of the evolution of the system.

It is a characteristic trait of our system that it is not closed, even when no photon is emitted. In fact, this enduring failure to observe a photon in the interval $[0,T]$ represents the continuum version of a series of non-detection measurements or negative result measurements. We have called them null measurements. A null measurement influences the system in just the same way as the usual positive result measurement. It would be therefore misleading to call it an 'interaction free measurement' (Dic81; Dic86; EV93). Perhaps it would be better called an 'energy exchange free measurement' (KBF98). Our setup may serve as a simple example in which the concept of null measurements leading to wave function collapses which result in level freezing can successfully be applied. Other types of null measurements which are performed on neutron spin (PNBR93) or on photon polarization (KWH$^+$95) are discussed in the literature.

Finally, we add that a full and detailed quantum optical treatment which is not based on a complex energy (as in Eq. (9.1)) leads to the same result. It is therefore possible to obtain for our system the quantum Zeno effect within a fully microphysical approach. Note that it was not necessary to refer to a collapse of the wave function. We will now compare these results with the usual phenomenological approach to the Zeno effect, which is based on repeated measurements.

9.3. Comparison with repeated projection measurements

Let us first consider a situation for which the setup of Itano *et al.* (see Sect. 4.1.2) may be regarded as an experimental realization. We have a system with two levels $|1\rangle$ and $|2\rangle$. We perform a sequence of strong measurements resulting in projections of the state vector (the continuous case will be discussed below). The two corresponding measurement results are R_1 and R_2. If R_i is measured the system is transferred to level $|i\rangle$. Accordingly the measurement of R_i is equivalent to the information that the system is immediately afterwards at level $|i\rangle$.

Let us assume that the first measurement of a sequence is performed at $t = 0$ (thus fixing the initial state) and then repeated N times with a (time) interval of τ between the consecutive measurements. τ is the relevant parameter. After the first repetition at time $t = \tau$ the system will still be found at the initial level with probability q or at the other level with probability $p = 1 - q$. After the time $T = N\tau$ the probability that the system has made a transition to the other level k times and has stayed at the initial level $N - k$ times is $P(k) = p^k q^{N-k}$. The probability that the system will be found finally at $t = T$ at the initial level, regardless of what have been the results of the measurements until then, is

$$P^Z(T) = \sum_{k \text{ even}} \binom{N}{k} p^k q^{N-k} . \tag{9.10}$$

For any fixed T the probability $P^Z(T)$ tends to 1 for $\tau \to 0$ (or $N \to \infty$).

For the experiment of the previous section the Rabi oscillation during the time τ fixes p and q to be

$$p = \sin^2 \Omega\tau, \quad q = \cos^2 \Omega\tau . \tag{9.11}$$

To compare with the sequential approach sketched above, we have to take into account that our system is peculiar in the following sense: the first measurement result R_1 is 'no emission of a photon'. This is a null result. We know that the system is at the initial level $|1\rangle$. There is no alternative measurement result R_2 because 'emission of a photon' is not connected with a projection onto $|i\rangle$, but indicates the end of the measurement series. Detection and emission of a photon are identified (see below). We know that all the measurements before have had the null result R_1. The related probability of permanent survival at level $|1\rangle$ is obtained as the first term ($k = 0$) of the sum (9.10):

$$P^Z_{ps}(T) = \cos^{2N}(\Omega\tau) . \tag{9.12}$$

We turn now to the case of very rapid repetitions with $\tau \ll \Omega^{-1}$. In this limit we may rewrite Eq. (9.12) without restriction of the time T of permanent survival as

$$P_{ps}^{Z}(T) = \exp(-\Omega^2 T\tau) . \tag{9.13}$$

Accordingly the result is that we can reach complete agreement with $P_{ps}(T)$ of Eq. (9.9) by identifying

$$\tau = \frac{2}{\gamma} . \tag{9.14}$$

The repetition time τ must be chosen as the inverse of half of the relaxation rate. The quantum optical treatment of Sect. 9.2 uniquely fixes in our comparison the otherwise unspecified parameter τ. The condition $T \gg \gamma^{-1}$ of Sect. 9.2 corresponds to the evident demand $T \gg \tau$. Perfect level freezing is obtained in the limit $\tau \propto \gamma^{-1} \to 0$.

It is interesting to see that because of Eq. (9.14) the repeated collapse to the ground state happens after time intervals τ which are completely fixed by the atomic lifetime. No reference to a photon detector is necessary. This demonstrates that the freezing at the level is independent of the detection or nondetection of the emitted photon in a real photon detector. Instead, it is the possibility of the irreversible decay into vacuum caused by the interaction with the environment which is responsible for the level freezing.

The fact that we are dealing with null measurements establishes another difference compared to the Itano *et al.* experiment. There the probability has been investigated of finding the initial state at time T regardless of the past history of the system in the interval $[0, T]$. This is to be described by $P^Z(T)$ given in Eq. (9.10). But, as has been pointed out in (NNPR96), this is not the genuine quantum Zeno effect according to Misra and Sudarshan, which is based on the idea of permanent survival as expressed by $P_{\text{ps}}^{Z}(T)$ of Eq. (9.12). This needs continuous observations such as can be found in our proposal.

9.4. Treatment within a complex Hamiltonian approach

The level freezing which we have predicted above is a realization of a genuinely continuous (strictly not sequential) measurement of energy, which has led to one uninterrupted null result of duration T. This represents a particular measurement readout $[E]$, namely $E(t) = E_1 = \text{const}$, with $0 \le t \le T$. Let us consider this situation in the framework of the RPI or complex Hamiltonian approach.

According to this approach a system with the Hamiltonian $H_0 + V$, subject to the continuous measurement of energy resulting in the readout $[E] = \{E(t)\}$, is described by the Schrödinger equation with the effective Hamiltonian containing an imaginary part (see Sect. 7.1):

$$H_{[E]} = H_0 + V - i\kappa(H_0 - E(t))^2, \tag{9.15}$$

where the constant κ characterizes the strength of the measurement. The inverse of this constant is a measure of the measurement fuzziness.

Returning to our passive Zeno setup, let us introduce a 2-level Hamiltonian H_0 with eigenvalues $E_i = \hbar\omega_i$ and rewrite our Hamiltonian H_γ, which describes the system without the driving field, in the form

$$H_\gamma = H_0 - i\hbar\gamma|2\rangle\langle 2| \,. \tag{9.16}$$

Then the complete Hamiltonian $H_\gamma + V$ used in Sect. 9.2 for the description of our system is identical to Eq. (9.15) provided that $E(t) \equiv E_1$ and the coefficient κ is expressed by the level width γ and the energy difference of the levels $\Delta E = \hbar(\omega_2 - \omega_1)$ as follows:

$$\kappa = \frac{\gamma}{\Delta E^2} \,.$$

Notice that the measurement readout $E(t) \equiv E_1$ means in the complex Hamiltonian approach that the system, according to the measurement, is staying at the level $|1\rangle$ all the time. Thus there is complete agreement between the results of the complex-Hamiltonian approach and our dynamical consideration of Sect. 9.2.

The so called level resolution time $T_{\rm lr} = 1/(\kappa\Delta E)$ is a measure of the weakness of the continuous measurement and of the resulting fuzziness of the readout. It is in our case $T_{\rm lr} = \gamma^{-1}$. The condition $T_l \ll \Omega^{-1}$ characterizes what is called the Zeno regime (strong influence of the measurement). This agrees with the condition used above in Sects. 9.2, 9.3. The comparison with the phenomenological complex Hamiltonian approach has enabled us to introduce the concept of the weakness of our continuous null measurement and to introduce a quantitative characteristic for this weakness. H_γ refers to the particular readout $[E] = E_1$. According to the complex Hamiltonian approach the probability of obtaining this $[E]$ agrees with $|a_1(T)|^2$ of Eq. (9.9).

9.5. Similar systems

Some systems resembling the one considered above have been discussed in the literature.

In the papers (Kra81; CHR95) continuous null measurements of transitions in a 2-level 'atom' ($|1\rangle$,$|2\rangle$) were considered. The atom occupies level $|1\rangle$ at the initial moment of time, its evolution without observation is described by the equation $|\dot\psi\rangle = -{\rm i}\beta\hat\sigma_1|\psi\rangle$, thus the atom oscillates between two levels with a frequency of β. The atom is exposed to a continuous observation by an apparatus indicating the transition from level $|1\rangle$ to level $|2\rangle$. The apparatus is coupled only to the state $|2\rangle$ of the atom by an interaction Hamiltonian of the form $|2\rangle\langle 2| \otimes H$, where H is a self-adjoint operator in the Hilbert space of the apparatus. This coupling is assumed to switch on a process in the apparatus resulting in the quick indication of the transition. Neglecting the inner dynamical evolution of the apparatus, the authors showed that the evolution of the system is inhibited if the apparatus is sufficiently fast. Their conclusion is that the Zeno effect may take place and that it can be considered as a consequence of purely dynamical unitary evolution of the combined system object–apparatus. The scheme of observation considered in (CHR95) is, however, rather abstract, and no experimental realization of this scheme has been discussed.

The 3-level scheme identical to the one considered by us has been mentioned in the work of Plenio *et al.* (PKT96) in connection with the Zeno effect, but with regard to a quite different aspect. Without any detailed consideration, it was noted that the oscillatory behavior of the probability P_1 is inhibited and replaced by an exponential one if level $|2\rangle$ spontaneously decays to another state during a sufficiently short time. This feature of the system was related to a finite width of the level $|2\rangle$, but the inhibition of oscillations was not connected with the quantum Zeno effect. On the contrary, the authors suggested to use the resulting behavior of level $|1\rangle$ as a model for a decaying system which shows a long enough non-exponential period. This 'artificial' decaying system had then to be measured repeatedly to demonstrate the Zeno effect.

Rabi oscillations with radiative damping have been studied in different contexts. We give an example: The state $|1\rangle$ is considered to be the ground state. A higher state $|2\rangle$ is assumed to be unstable. The system decays back to the state $|1\rangle$ with the emission of a photon. Investigation of such a system produces the well known result that Rabi oscillations are damped by radiative decay. If the decay rate of level $|2\rangle$ is much greater than the Rabi frequency the probability of finding the system at the level $|1\rangle$ remains close to unity at all times. In this case (in contrast to our 3-level system) there are simultaneously two mechanisms of 'freezing' the system at the level $|1\rangle$: the fast transition from level $|2\rangle$ back to level $|1\rangle$ and, in addition, Zeno-like inhibition of Rabi oscillations connected with broadening of the level $|2\rangle$. Stimulated and spontaneous emission happen together. It is difficult to separate the two mechanisms. We cannot interpret the resulting behavior as Zeno effect, because the two possible observational results — that a photon is or is not emitted — are both related to the same information, namely that the system is transferred to level $|1\rangle$. This example shows that not any level 'freezing' may be taken as an indication for the presence of a Zeno type of measurement influence. In fact the analysis must be carried out the other way round: If a situation can be interpreted as a measurement of Zeno type (including null measurements) then level freezing is to be expected.

9.6. Conclusions

We have shown that the freezing at the initial level (the inhibition of a transition) caused by the possibility of spontaneous decay of the other level can indeed be understood as a Zeno effect, because it goes back to the measurement of the system. This is supported by a comparison with the phenomenological scheme based on repeated projection measurements and the phenomenologically described continuous energy measurement in the Zeno regime. After appropriate adjustment of parameters they all agree with the result of the quantum optical calculation.

The Zeno effect as described above is a particular quantum phenomenon related to quantum measurements which may occur in many different physical situations. To refer to it is of great predictive power: before any microphysical calculation involved has been performed it must be expected that freezing at a level (or at least the slowing down of a transition) will happen if measurement influences the system.

The advantage of the particular system presented above is that it is easier to analyse than the standard example of Itano *et al.* (Sect. 4.1.2) and shows, nevertheless, to a certain extent a counter-intuitive behavior. At the same time it is a very simple example of a null measurement, which is sometimes not quite correctly called an interaction-free measurement. Potentially the most interesting aspect of the system may be that it represents a truly continuous measurement, which can easily be handled, and which we have discussed above only in its Zeno limit.

The problem considered here is closely connected with a more complicated problem of the quantum Zeno effect in the decaying system. Originally the Zeno effect was discussed in the case of a spontaneously decaying system (MS77; CSM77; Per80), and preventing (or slowing down of) the decay was predicted as a consequence of the continuous observation of the decay. Later on it was argued (GOWR79; Coo88; Per85; IHBW90) that the Zeno effect cannot arise in spontaneous decay.

It is natural to apply to the decaying system the same consideration which has been used here in the case of the Rabi oscillation. Namely, if the target level of the decaying system is unstable, the very capability of this level to decay may be considered as a continuous observation of the primary decay, and therefore could lead to preventing the primary decay or at least to its slowing down.

In (Men99; Pan99) this idea has been confirmed. It has been shown, under certain assumptions, that the decay is slowed down if the target level is unstable. The conclusion made in (Men99) was that under certain conditions the instability of the target level of the decaying system may be considered as a continuous measurement leading to the Zeno effect, but the precise formulation of the conditions is yet to be discussed in more detail in the following sense.

Observation of the decay is characterized by the time of observation and by the energy of the decay products. Simple statements that the decay 'is observed' or 'is not observed' are hardly completely adequate. One needs quantitative characteristics of the observation in order to accurately define what may be called the Zeno effect for the decaying system and then prove the existence of this effect. An attempt of such analysis has been carried out in (Pan99) where, in particular, the comparison was done with the slowing down of the Rabi oscillations by the instability of the target level (a subject of the present chapter).

Part IV

Conceptual Problems and Prospects

Chapter 10

Restricted Path Integrals and the Philosophy of Quantum Mechanics

Quantum mechanics and particularly the quantum theory of measurements, was always the arena of keen discussions about conceptual questions. In the present chapter we concern ourselves with some of the conceptual aspects of quantum mechanics which result from the theory of continuous quantum measurements, or obtain a new insight from the point of view of this theory formulated in terms of restricted path integrals (RPI) or quantum corridors.

We argue that: 1) quantum mechanics becomes conceptually closed when formulated in terms of Feynman path integrals including RPI; 2) the RPI approach to continuous measurements reveals the intriguing dynamical role of information; 3) with this approach, open systems may be considered to be primary concept as compared to closed systems. Finally we speculate on the emergence of classical features in the quantum world, on the possible role of consciousness in this, and on imaginary potential as presenting this emergence.

10.1. Conceptually closed quantum mechanics

The description of continuous quantum measurements by restricted path integrals (RPI) may be justified in different ways. The most direct way is an analysis of a composite system containing both the measured system and its environment (the measuring device). Alternatively, one can consider a series of instantaneous measurements with the help of von Neumann's projection postulate and then go over to the continuous measurement as a limit of small time intervals between the instantaneous measurements.

It is very interesting, however, that in the framework of the Feynman version of quantum mechanics the RPI approach needs no justification at all. This approach is natural and self-consistent in this framework. This is why Feynman was able to formulate the idea of the RPI approach as a short remark in his paper (Fey48). Moreover, the RPI approach, naturally following from Feynman's quantum mechanics, is more general than what can be obtained in the limit of a series of instantaneous measurements. It describes a wider class of continuous and continual (i.e., protracted in time and space) measurements than those derivable as limits of some or other repeated measurements (see Sect. 5.3).

The reason why the RPI approach follows from the path integral version of quantum mechanics is that the concept of *probability amplitude* is used in a much more comprehensive way in this version. In particular, the amplitude

$$A[p,q] = e^{\frac{i}{\hbar}\int_0^T (p\dot{q} - H(p,q,t))} \tag{10.1}$$

is introduced and physically interpreted as a probability amplitude for the system to propagate along a definite path in the phase space. If this is accepted, then the usual quantum mechanical rules for amplitudes determine the amplitudes for more complicated events, in particular, for propagation of the system between two points of the configuration space. If the system is closed, and therefore nothing is known about the path along which it propagates, all amplitudes of the form (10.1) have to be summed together, leading to the conventional Feynman integral. If a continuous measurement takes place (so that the system is open), one has to keep in mind that the measurement supplies some information about the evolution of the system. In summing up the amplitudes, Eq. (10.1), this information must be taken into account.

If the information given by the measurement can be expressed by a weight functional, then summation of amplitudes takes the form of a weighted path integral. Hence, instead of directly postulating partial propagators we can *derive* them from the more basic postulates of the path integral version of quantum mechanics.

This is both interesting and unexpected. It is commonly believed that quantum mechanics is not closed, since it does not include any theory of measurements. A theory of measurements (for example, von Neumann's projection postulate) is customarily appended as a necessary counterpart forming, together with quantum mechanics, a closed theory. However, this proves to be unnecessary. As argued above, the path integral formulation of quantum mechanics also includes the RPI theory of continuous measurements. A theory of instantaneous measurements (including von Neumann's postulate) may be then obtained as a limit (Cal95).

We see, therefore, that the seeming necessity of postulating a theory of measurement independently of quantum mechanics is only a consequence of over-idealization. The origin of this necessity is in treating a measurement as an instantaneous act. An instantaneous measurement appears to be external with respect to quantum mechanical (Schrödinger) evolution and to need special postulates. The situation, however, is radically different if the measurement is considered as a temporarily extended process and quantum mechanics is accepted in the path integral form. The measurement may then be described together with the quantum mechanical evolution in a non-separable way (Men88b). The mathematical apparatus describing both counterparts of this unity is given by restricted path integrals.

10.2. The information approach to continuous measurements

Restricted path integrals describe the influence of the measuring environment on the open (measured) quantum system without an explicit model of the environment. Instead, the RPI approach needs only very general characteristics of the environmental state. Namely, it is necessary to know what information about the evolution of the system is recorded in the state of the environment (as a measurement readout). This information determines what weight functional has to be used in the path integral. Having a restricted path integral, we can describe correctly both the probability distribution of measurement readouts and the final state of the measured system.

Thus the influence of the environment on the system of interest may be given in terms of information. Therefore the RPI approach is in fact an *information approach* to the theory of continuous quantum measurements. This indicates the fundamental character of the approach.

The information approach in the quantum theory of measurement is not novel. The very first quantum theory of (instantaneous) measurements based on the von Neumann reduction postulate may be considered as an example of an information approach. In this theory both the probability distribution for measurement outputs and the final state of the system may be found if we know what information the measurement supplies. For example, for the measurement of an observable with a discrete spectrum we need to know what eigenvalue is obtained as the measurement output. Given this information we can determine, with the help of the corresponding projector, both the probability of the measurement output and the final state of the system.

In the RPI theory of continuous quantum measurements the information principle obtains one more realization which has a rich structure and wide range of applications. RPI theory is, in contrast to the von Neumann postulate, a natural part of (the path integral version of) quantum mechanics.

Because of the great importance of these questions, let us consider them from a slightly different point of view.

The measured system is an open one (influenced, from outside, by its environment). Therefore it is only a subsystem of a larger system. Usually the state of the subsystem may be represented by a density matrix, not by a state vector (a wave function), see Appendix A1. It is possible to represent the state of a subsystem by a vector if the state of its environment is given.

In the RPI approach a continuously measured (and therefore open) system is described by a vector (not a density matrix), but in terms referring only to this system. The environment is not explicitly present in this description. How is this possible? How can we fix the state vector of a subsystem not referring to the state of its environment, in terms of the subsystem itself? This turns out to be possible because, instead of the state of the environment, we fix *information* about the subsystem recorded in the state of the environment.

This is a very deep phenomenon, yet to be thought about and investigated. It may be formulated in the following way. Not knowing the state of the environment, and even knowing nothing about its structure and the interaction with our system, we can take into account the influence of the environment onto the system. To this end we need know only *what information about the system is recorded in the environment.* This is enough to calculate the influence, and this is done in the RPI approach.

Why is this possible? The answer may be only that this is a law of nature. This law is in fact comprehended in the Feynman path integral form of quantum mechanics as 'potential capability' and realized in the RPI theory of continuous measurements.

10.3. Which is primary: closed or open systems?

What has been said in Sect. 10.2 radically changes the status of the question of which is primary: closed or open systems? It is conventional to consider a closed system as primary and to derive open systems as subsystems of closed ones. This is natural, convenient, and may even seem necessary. However, we see now that an open system may be described in a conceptually closed way, so why cannot it be primary?

This alternative approach may be set up as follows. Let us consider an open system which interests us. Its description is closed. Even (unavoidable for an open system) external influences can be taken into account in terms of the system itself. To this end we need to assume something about its environment, and it will be the character of information about the system recorded in the environment. Then, specializing the concrete contents of the information, we know, by RPI technique, what the behavior of the system is. By changing the hypothesis about the recorded information we may treat systems which have different environments but are otherwise the same.

If we really do need to describe a part of the environment explicitly, we go over to a larger system containing the original one as a subsystem.

However, this system will again be open, and again we can take into account the influence from outside in the framework of the information approach, i.e., in terms of the system itself.

In this manner we may consider open, and only open, systems, never speaking about closed systems. If we need (and of course this is often convenient) we can describe a closed system as a limit of an open system when the influence of the environment is negligible. Technically this will be achieved by widening quantum corridors which restrict path integrals or by diminishing imaginary terms in the Hamiltonian.

This approach which deals only with open systems seems even more realistic than the conventional approach because of two facts: first, we never meet strictly closed systems, and second, for some open systems we know too little about their environments, so that it is advantageous to describe such a system in its own terms and not in terms of its environment.

There is one more argument, although it may be too abstract: there is, in fact, no strictly closed systems (except the whole Universe), so that theory of closed systems is in fact empty. This of course may sound too speculative, but in Sect. 10.4 a somewhat similar, but more concrete, argument will be formulated.

A theory of open systems in terms of themselves may be called a *closed theory of open systems*. It is usually much more economical (compared to the consideration of open systems as subsystems) because it does not require models of environments. As a rule this is important, since the environment (the reservoir) is often a complicated system. Taking its influence implicitly within the RPI (or complex-Hamiltonian) approach, we may have a great advantage, because this will be done in the simplest way: only those few features of the sophisticated environment will be included which are actually material to the system of interest.

The preceding argument leads us to the next question: *Which is primary: determinate or stochastic evolution?*

The answer to this question evidently follows from the preceding question. Conventionally deterministic systems are considered as primary. Stochastic properties are derived from incomplete knowledge and therefore are considered as secondary. Of course, in quantum mechanics we call deterministic the evolution given by a wave function (state vector) satisfying Schrödinger equation. If we use (as suggested above) the closed theory of an open system as a primary theory, then the evolution in this primary theory is stochastic.

Indeed, the evolution operator of an open (a continuously measured) system is described by a RPI and depends on some function (or functions) interpreted as a measurement readout. This function is random, and its probability distribution can be calculated. Alternatively, the wave function (the state vector) of the system may be found as a solution of the complex-Hamiltonian Schrödinger equation also depending on the measurement readout. Equivalently one may use a stochastic Schrödinger equation. In any case random functions enter the theory. If we consider open systems as primary, stochastic evolution becomes primary too. Deterministic evolution (the usual Schrödinger equation) may be then obtained as a limit of the complex-Hamiltonian or stochastic Schrödinger equation.

10.4. The emergence of classical features in the quantum world

We stand now before the hardest and most intriguing question, how does the classical appearance emerge from an originally quantum word.[1] In a sense, this is now a very simple and clear question. From another point of view it it hard, not yet solved, and may even be insoluble. Nevertheless, we shall try to speculate on the topic.

Let us first formulate the question in the simplest possible form. If a particle can be localized at one of two points at a macroscopic distance from each other, then in the quantum world a superposition of these two localized states may also exist. However, in our (classical) real world we always see the particle at one of these two points. Why at only one? How does the selection between the two alternative localizations occur?

The simple answer to the question is decoherence. The superposition, if being prepared, will very quickly decohere, because of interaction with the environment, into a mixture of two localized states. The coefficients of the mixture give the probabilities of the alternatives. The necessity of the process of decoherence may be easily proved (see Sect. 2.1). This solves all practical questions, allows one to analyse complicated measurements, and to make predictions. However, conceptual questions still remain.

Briefly speaking, we may easily understand how the exclusion of the environment converts a superposition into a mixture, but this does not explain how one of the alternatives may be selected. One of the formulations of the so called 'measurement problem' is how selection of a single alternative may be explained (in this connection see Sect. 2.2.2). Let us give also one more formulation connected with this.

We saw in Sect. 2.3 that the attempt to explain how the measurement (decoherence) occurs leads finally to one more measurement of another system (the environment with respect to the first system). Of course, this new measurement can also be explained in an analogous way, but this will finish with one more measurement of the environment with respect

[1]We will not discuss the alternative, that our world is not originally quantum, because its quantum character on the fundamental level is strongly supported by physical experiments (this is not a final proof of course).

to the environment, and so on. The chain is infinite, up to the cells in the observer's eye or even further, to cells in his brain. The necessity of somehow breaking this infinite chain, introducing the frontier between the system and observer, is the essence (or at least one of the main aspects) of the measurement problem in quantum mechanics.

Just in this point there is a possibility that may be suggested as a possible solution. Returning to Sects. 10.1, 10.2, 10.3, we see that in the closed theory of open systems (based on RPI) there is no infinite chain. The object of such a theory is always a finite open system. It is subject to measurements by the environment, and the measurement output is an inherent element of the theory. We can, of course, extend the consideration to include a part of the environment in it; however, again we shall have some measurement procedure as a fundamental element of the theory. We need not break the chain to obtain the measurement because the measurement is included in the theory from the very beginning. This may be a one more argument for the theory of open systems as the fundamental, primary theory.

The final conclusion is that the measurement problem does not arise if we accept the theory of open (continuously measured) system as the primary theory, considering closed systems only as a limiting case of open systems. The problem arises as a result of over-idealization. We are too proud of our mind if we hope to formulate a theory of closed systems, because there is in fact only one closed system, the Universe.

Nevertheless, one may wish for more than this. Even though any real system is open, a scientist (clearly too proud of his mind) may wish to understand where the infinite chain of measurements of measurements lead. The answer seems unambiguous. We have seen that a chain of measurement after measurement cannot terminate at any physical system. However, we know the point at which there is only one alternative, i.e., the measurement is already over. This is consciousness.

The chain therefore leads to consciousness. This is why the 'measurement problem' is not, to our mind, a physical problem. It is a metaphysical (philosophical) problem and can be solved (if at all) only with the inclusion of such a concept as consciousness (see in this connection (Wig83)). Being a physicist, one may not wish to go as far as this. Nevertheless, there were many attempts to realize this ambitious program, most of them based on Everett's many worlds interpretation of quantum mechanics (see Sect. 2.2.2).

One of the most interesting conceptual systems for the realization of this program was built by Euan Squires (Squ94) (see also (Loc96)). According to that, consciousness is a window opening into the quantum world. It does not allow one to see the whole quantum world, but only a single fragment of it. In terms of the theory of quantum measurement, only one out of all possible alternatives. According to Everett's interpretation, only one of his many worlds. This is how classical reality may emerge in the quantum world.

Can the theory of restricted path integrals contribute to this ambitious program? It seems that it gives an instrument which may at least help to formulate the principal point of the program in an elegant way. This is an imaginary potential arising in the complex-Hamiltonian approach (more generally, a weight functional in RPI).

We have seen in this book (Chapter 5) that the influence of the measurement on the measured system is represented by an imaginary term in the Hamiltonian depending on the measurement readout. This means that the selection of a single alternative is described by an imaginary potential. If we accept (as has been advocated above) that this selection is the work of consciousness, we are forced to conclude that the *work of consciousness is described by the imaginary potential in the Hamiltonian.*

The final scheme then looks as follows. A physical system is described by a real Hamiltonian. Inclusion of an imaginary potential of the specified form allows one to describe the influence of the environment in which information about the system is recorded. The system is then treated as open. The concrete form of the imaginary potential describes the selection of a single record from all possible alternatives. If we ask how the selection is made, we are forced to answer that this is the work of consciousness. There is no necessity, if we are inside physics, to ask this. Then we use the imaginary potential as an efficient mathematical instrument for the description of decoherence. If, however, we are curious to 'understand' it deeper, the only possible answer is that the imaginary potential represents consciousness.

It is interesting to compare the roles played by real and imaginary potentials in the Hamiltonian. The real potential also represents the influence of external systems. This influence is summed up as the action of external classical fields. This means that the fields are assumed fixed, the back influence of our system on the origins of the fields being negligible. In this situation no information about our system may be recorded in those external systems which create real potential. As a result the influence of these systems does not induce decoherence.

The situation is quite different in the case of an imaginary potential. It also describes the influence of external systems (the environment), but a qualitatively different influence. It is originated by such systems which can suffer back influence, and therefore may change their state in accord with the state of our system. Therefore information about the system's state may be recorded in such systems. This results in decoherence of the system represented by the imaginary potential.

What is the role or nature of imaginary potential? In this book we have advocated it as a convenient instrument for the phenomenological description of quantum measurements providing its *selective description*. According to what has been said above about the role of consciousness in the selection of alternatives, or imaginary potential describes (also phenomenologically) the operation of consciousness.

Chapter 11

Conclusion: New Prospects

We have considered in the present book the phenomenon of decoherence as a realization of a quantum measurement and *vice versa*: continuous quantum measurements as a phenomenological description of gradual decoherence. Theory of continuous quantum measurements developed on the basis of restricted path integrals (RPI) became thus the phenomenological theory of gradual decoherence. The RPI formalism has been demonstrated to be conceptually significant: it shows how information about the state of a quantum system governs its dynamics.

The latter makes RPI an important mathematical and conceptual apparatus in quantum theory as a whole, beyond the range of those non-relativistic problems which have been discussed in the present book. Some relativistic problems have already been considered by the RPI method in the previous book (Men93), among them the measurability of electromagnetic and gravitational fields (see also (vBM94)). Other relativistic problems (the Unruh effect and quantum effects in black holes) were explored in (Men98c). However, the work on relativistic applications of the method is now in its initial stage.

Phenomena of the type of decoherence are usually not considered in relativistic quantum theory. Moreover, unitarity (and therefore the absence of decoherence) is conventionally accepted as one of the main principles of quantum field theory (QFT). However, this is nothing but an over-idealization leading, by the way, to the famous problem of QFT, infinities. Efficient methods are elaborated for overcoming this difficulty by renormalization. However, this way is to some degree artificial. The very possibility of removing infinities by renormalization hints that they may be avoided from the very beginning in a more adequate theory.

The solution of this problem has for many years been sought in the theory of superstrings, which has become, in fact, a complicated branch of mathematics rather than physics. To our mind, it is much more natural to expect that the solution will be found in a quite different direction, in non-unitary theory taking decoherence into account. This would be in the spirit of the idea, formulated in Sect. 10.3, that open systems may be considered as primary systems. In Sect. 6.2 it has been demonstrated how this approach overcomes well known difficulties in the definition of a path integral. The same may be expected in relativistic quantum theory. The relativistic modification of the RPI formalism must be very useful in formulating such a theory.

This may be a very interesting direction of investigation in quantum theory. Mathematical aspects of this area are yet to be elaborated. It is clear that they will be quite different from what we have used in the non-relativistic field. One may also hope that qualitatively new physical ideas might be found in this way. It is especially interesting to apply this direction of investigation to quantum gravity, where new points of view are evidently necessary.

The Formalism of Quantum Mechanics

In this Appendix we present only those concepts of quantum mechanics which are of great importance for the book and are not commonly well known: density matrices, the mathematical background of the Heisenberg uncertainty relations, and Feynman path integrals.

A1. The density matrix

The density matrix is used for the representation of a wider class of states of a quantum system than pure states represented by state vectors $|\psi\rangle$. Yet all essentially quantum phenomena may be described in the language of pure states (state vectors) because, as we shall see below, a density matrix contains both quantum and classical elements (respectively state vectors and probabilities).

A1.1. SUBSYSTEM STATE

To arrive at the concept of a density matrix we shall consider the system of interest as a subsystem of a larger system; but first we introduce the trace of an operator.

Trace of an operator
An arbitrary operator B may be represented by its matrix elements:

$$B = \{B_{ij} = \langle\psi_i|B|\psi_j\rangle\}.$$

The trace of the operator is defined as the sum of its diagonal elements:

$$\mathrm{tr}B = \sum_i B_{ii} = \sum_i \langle\psi_i|B|\psi_i\rangle. \tag{A.1}$$

Making use of the property of any complete orthonormal basis $|\psi_i\rangle$,

$$\sum_i |\psi_i\rangle\langle\psi_i| = \mathbf{1},$$

one may show that the value (A.1) does not depend on the choice of the basis. It is easy to derive the following very simple rule for the evaluation of an operator of the special form, $B = |\varphi\rangle\langle\chi|$:

$$\mathrm{tr}(|\varphi\rangle\langle\chi|) = \langle\chi|\varphi\rangle,$$

and then, applying the linearity of the operation of taking the trace,

$$\mathrm{tr}\left(\sum_\alpha a_\alpha |\varphi_\alpha\rangle\langle\chi_\alpha|\right) = \sum_\alpha a_\alpha \langle\chi_\alpha|\varphi_\alpha\rangle.$$

Analogously, using the definition of the trace, we can show that cycling of the product of operators may be made under the sign of trace:

$$\mathrm{tr}(ABC\ldots Z) = \mathrm{tr}(BC\ldots ZA),$$

which implies the following important relation:

$$\mathrm{tr}(U^{-1}AU) = \mathrm{tr}(A).$$

We can now rewrite the mean value of an operator A in a state $|\psi_i\rangle$

$$\overline{A} = \langle\psi|A|\psi\rangle = \langle A\rangle$$

as follows:

$$\overline{A} = \langle A\rangle_\rho = \mathrm{tr}(A\rho) \tag{A.2}$$

where the density matrix for the pure state $|\psi\rangle$ is defined as

$$\rho = |\psi\rangle\langle\psi|.$$

Reduced density matrix

Let now $\mathcal{S}$ be a subsystem of the compound system $\mathcal{S}+\mathcal{E}$, the of basis $\mathcal{S}$ being $\{|\psi_i\rangle\}$ and the basis of $\mathcal{E}$ being $\{|\Phi_i\rangle\}$. Then the basis of the compound system $\mathcal{S}+\mathcal{E}$ is $\{|\psi_i\rangle|\Phi_\alpha\rangle\}$ and a general state of $\mathcal{S}+\mathcal{E}$ has the form

$$|\Psi\rangle = \sum_{i\alpha} c_{i\alpha}\, |\psi_i\rangle|\Phi_\alpha\rangle$$

According to Eq. (A.2), the mean value of an operator A in this state is

$$\langle A\rangle_\rho = \langle\Psi|A|\Psi\rangle = \mathrm{tr}_{\mathcal{S}\mathcal{E}}(A|\Psi\rangle\langle\Psi|)$$

If, however, the operator A acts only on the subsystem $\mathcal{S}$, then the trace over $\mathcal{E}$ may be referred only to the second factor $|\Psi\rangle\langle\Psi|$:

$$\langle A\rangle_\rho = \mathrm{tr}_{\mathcal{S}}(A\rho),$$

where we have denoted

$$\rho = \mathrm{tr}_{\mathcal{E}}(|\Psi\rangle\langle\Psi|).$$

The operator ρ (acting in the subsystem $\mathcal{S}$) is called a *reduced density matrix* of this subsystem $\mathcal{S}$ and is equal to

$$\rho = \mathrm{tr}_{\mathcal{E}}(|\Psi\rangle\langle\Psi|) = \sum_{\alpha}\sum_{ij} c_{i\alpha}c^*_{j\alpha}\,|\psi_i\rangle\langle\psi_j| \tag{A.3}$$

The state of the subsystem cannot (in the general case) be described by a state vector. Instead, it should be described by a density matrix. Such a state is called *mixed.*

A1.2. INTERPRETATION OF A DENSITY MATRIX

Structure of the reduced density matrix

It is easily seen that the density matrix (A.3) may be represented in the following form:

$$\rho = \sum_{\alpha} |\varphi_\alpha\rangle\langle\varphi_\alpha|,$$

where

$$|\varphi_\alpha\rangle = \sum_i c_{i\alpha}|\psi_i\rangle.$$

Going over to the normalized vectors

$$\begin{aligned} |\psi_\alpha\rangle &= \frac{|\varphi_\alpha\rangle}{\sqrt{p_\alpha}}, \\ p_\alpha &= ||\varphi_\alpha||^2 = \sum_i |c_{i\alpha}|^2 \end{aligned}$$

we have

$$\rho = \sum_{\alpha} p_\alpha|\psi_\alpha\rangle\langle\psi_\alpha| \tag{A.4}$$

(notice that in the general case the states $|\psi_\alpha\rangle$ are not mutually orthogonal).

From now on we shall denote by 'tr' the trace over the system $\mathcal{S}$. The following properties of the density matrix can be easily proved:

$$\mathrm{tr}\rho = 1, \quad \rho^\dagger = \rho, \quad \rho \geq 0 \tag{A.5}$$

The first property (of unit trace) is equivalent to

$$\sum_{\alpha} p_\alpha = 1$$

and follows from the fact that the original state $|\Psi\rangle$ was normalized:

$$\sum_{i\alpha} |c_{i\alpha}|^2 = 1.$$

In the special case where the density matrix represents a pure state, it is an idempotent operator, or a projector:

$$\rho = |\psi\rangle\langle\psi| \quad \Rightarrow \quad \rho^2 = \rho$$

Probability interpretation
The properties

$$p_\alpha \geq 0, \quad \sum_\alpha p_\alpha = 1$$

hint that the numbers p_α might be interpreted in the purely classical sense as probabilities of alternative (i.e., excluding each other) events. This guess is valid. The state ρ may be interpreted as a state with incomplete knowledge. We know that the system $\mathcal{S}$ may be in one of the pure states $|\psi_\alpha\rangle$, but do not know in which of these states. Instead, we know only the probability p_α of each of them. This is sometimes called the *ignorance interpretation.*

This interpretation is in fact a hypothesis, and it must be confirmed. The confirmation lies in the fact of the calculation of the mean value fulfilled on the basis of this hypothesis giving the correct result. Indeed, starting with our interpretation, we have to calculate the mean value of an operator A in each of the pure states $|\psi_\alpha\rangle$ and find the weighted average with weights p_α. This gives the correct result:

$$\sum_\alpha p_\alpha \langle A\rangle_\alpha = \sum_\alpha p_\alpha \langle\psi_\alpha|A|\psi_\alpha\rangle = \mathrm{tr}(A\rho) = \langle R\rangle_\rho .$$

A1.3. DECOMPOSITION OF A DENSITY MATRIX

Non-orthogonal set of vectors
Eq. (A.4) supplies a decomposition of a density matrix into a sum (with positive coefficients) of density matrices of pure states. If we have some density matrix, i.e., an operator ρ with the properties (A.5), it may always be decomposed in this way. If we do not require that the vectors $|\psi_\alpha\rangle$ be mutually orthogonal, decomposition is ambiguous. This may be seen from such a simple counter-example referring to a two-dimensional space:

$$p|1\rangle\langle 1| + 1|2\rangle\langle 2| = |\varphi_1\rangle\langle\varphi_1| + |\varphi_2\rangle\langle\varphi_2|$$

if

$$\begin{pmatrix} |\varphi_1\rangle \\ |\varphi_2\rangle \end{pmatrix} = \begin{bmatrix} \sqrt{p}\cos\chi & \sqrt{q}\sin\chi \\ -\sqrt{p}\sin\chi & \sqrt{q}\cos\chi \end{bmatrix} \begin{pmatrix} |1\rangle \\ |2\rangle \end{pmatrix}$$

with arbitrary χ.

Despite this ambiguity, the interpretation formulated above is valid for each of the decompositions in the sense that the calculation of mean values of operators may be achieved with the help of any of the decompositions.

We have to make an important remark in this connection. It is stated in the probability interpretation of the decomposition (A.4) that p_α is the probability that the system *is* in the state $|\psi_\alpha\rangle$. This statement, however, should not be confused with the answer to the question: what is the probability that the system *will be found* (by the corresponding measurement) in the state $|\psi_\alpha\rangle$. Naïvely one may answer this question with the same number p_α. In actuality the probability is larger.

Let us first prove this and then explain the apparent paradox. How can we answer the formulated question? We have to perform the measurement with two alternative results corresponding to the two (mutually complementary) projectors:

$$P = |\psi_\alpha\rangle\langle\psi_\alpha|, \quad Q = \mathbf{1} - P.$$

The measurement result corresponding to the projector P will mean the answer 'yes' to the above question. The probability of this result may be calculated according to the usual quantum mechanical rules (Sect. 2.3):

$$\mathrm{Prob}(P) = \mathrm{tr}(\rho\, P) = \sum_{\alpha'} p_{\alpha'} |\langle\psi_\alpha|\psi_{\alpha'}\rangle|^2. \tag{A.6}$$

The r.h.s. may be rewritten as

$$p_\alpha + \sum_{\alpha'\neq\alpha} p_{\alpha'} |\langle\psi_\alpha|\psi_{\alpha'}\rangle|^2,$$

which is generally larger than p_α.

The reason for this purely quantum paradox is that *to be* in some pure state and *to be found* in it are not the same things. If the system is in a state $|\psi_{\alpha'}\rangle$ not identical to $|\psi_\alpha\rangle$, it may be found (by the above measurement) in the state $|\psi_\alpha\rangle$ with probability $|\langle\psi_\alpha|\psi_{\alpha'}\rangle|^2$. Therefore, even if we know that the system is in $|\psi_\alpha\rangle$ with probability p_α, the probability that the system will be found in $|\psi_\alpha\rangle$ is given by Eq. (A.6).

Orthogonal set of vectors (eigenvectors)

Among all the decompositions of the density matrix there is only one decomposition in the orthogonal set of pure states (up to degeneracy, as we shall see). The corresponding orthogonal states are eigenstates of the operator ρ. Denoting them $|\psi_n\rangle$, we have

$$\rho\,|\psi_n\rangle = p_n|\psi_n\rangle.$$

From the properties (A.5) of the density matrix the following properties of its eigenvalues follow:

$$p_n \geq 0, \quad \sum_n p_n = 1$$

This gives, in fact, a spectral decomposition of the density matrix:

$$\rho = \sum_n p_n \, |\psi_n\rangle\langle\psi_n|. \tag{A.7}$$

The vectors $|\psi_n\rangle$ are automatically orthogonal if the corresponding eigenvalues are different and may be chosen orthogonal for coinciding eigenvalues (in the case of degeneracy). In any case (for both a degenerate and a non-degenerate spectrum) the decomposition into the orthogonal projectors on subspaces of different eigenvalues is unique:

$$\rho = \sum_N p_N \, P_N, \quad P_N P_{N'} = \delta_{NN'} P_N, \quad \sum_N P_N = 1$$

In the case of orthogonal decomposition there is no paradox. The probability of finding the pure state $|\psi_n\rangle$ in the mixed state ρ is equal to the corresponding coefficient p_n of the decomposition:

$$\mathrm{Prob}(|\psi_n\rangle\langle\psi_n|) = \mathrm{tr}(\rho \, P_n) = p_n$$

The spectrum of the density matrix may be continuous. In this case the sum in the decomposition must be replaced by the corresponding integral:

$$\rho = \int p(c) \, P(c) \, dc.$$

The orthogonality of projectors has now the form

$$P(\mathcal{A}) \, P(\mathcal{B}) = 0 \quad \text{for} \quad \mathcal{A} \cap \mathcal{B} = \emptyset,$$

where

$$P(\mathcal{A}) = \int_{\mathcal{A}} p(c) \, P(c) \, dc.$$

A2. Uncertainty relations

Let us derive the uncertainty relations (UR) for two Hermitian operators A, B. Form an operator $M = A - i\lambda B$, where λ is a real number, and notice that the operator $M^\dagger M$ is positive,

$$M^\dagger M \geq 0.$$

This means that its mean value in an arbitrary state $|\psi\rangle$ is non-negative,

$$\langle M^\dagger M \rangle = \langle\psi|M^\dagger M|\psi\rangle \geq 0 \tag{A.8}$$

(the latter follows from the fact that the l.h.s. is a norm of the state $M|\psi\rangle$).

Expressing $M^\dagger M$ in terms of A and B, we have from Eq. (A.8)

$$\langle \lambda^2 B^2 + \lambda C + A^2 \rangle \geq 0,$$

where

$$C = -i[A, B].$$

Owing to the linearity of the operation of averaging,

$$\lambda^2 \langle B^2 \rangle + \lambda \langle C \rangle + \langle A^2 \rangle \geq 0.$$

This inequality has to be valid for arbitrary λ, hence

$$\langle A^2 \rangle \langle B^2 \rangle \geq \frac{1}{4} \langle C \rangle^2.$$

Going over to the same operators by shifted by c-number terms, we have

$$\langle (A-a)^2 \rangle \langle (B-b)^2 \rangle \geq \frac{1}{4} \langle C \rangle^2.$$

At last, taking

$$a = \langle A \rangle, \quad b = \langle B \rangle,$$

and denoting

$$\langle (A - \langle A \rangle)^2 \rangle = \langle A^2 \rangle - \langle A \rangle^2 = \Delta A$$

we have the UR for the operators A, B:

$$\Delta A \Delta B \geq \frac{1}{2} |\langle C \rangle|.$$

In the case of the coordinate and linear momentum this gives the well known UR for them.

Taking the function $f(a) = \langle (A-a)^2 \rangle$ and looking for its minimum we find that it is achieved for $a = \langle A \rangle$. Therefore,

$$\langle (A - \langle A \rangle)^2 \rangle = \min_a \langle (A-a)^2 \rangle.$$

In the preceding derivation we have made use of the mean value defined by averaging over a pure state:

$$\langle A \rangle = \langle \psi | A | \psi \rangle.$$

However, the same is also valid for the mean value over a mixed state:

$$\langle A \rangle_\rho = \mathrm{tr}(A \rho).$$

Indeed, the operation of averaging is linear in this case too, and a positive operator has a non-negative mean value in an arbitrary mixed state (this follows from the decomposition (A.7) or (A.4)).

Therefore, all the preceding formulas including the UR are valid for the mean value $\langle\,\rangle_\rho$ as well as for $\langle\,\rangle$.

A3. Technics of path integrals

Here we present briefly some technical aspects of path integrals. The details may be found in Feynman's book (FH65), in (IZ80; Sch81; Kle90), and also in the previous book of the present author (Men93).

A3.1. PROPAGATORS AND PATH INTEGRALS

The Feynman path integral

$$U(t'', q''|t', q') = \int \mathrm{d}[q] \exp\left(\frac{i}{\hbar} S[q]\right) \tag{A.9}$$

is understood as an integral over all paths

$$[q] = \{q(t)|t' \le t \le t''\}, \quad q(t') = q', q(t'') = q'' \tag{A.10}$$

between the given points q', q'' in the configuration space (usually multi-dimensional) of a quantum system. Here S is an action of the system

$$S[q] = \int_{t'}^{t''} L(q, \dot{q}, t)\, \mathrm{d}t$$

expressed through its *Lagrangian* L, which in turn is related to the Hamiltonian in the usual way:

$$H(p, q, t) = p\dot{q} - L(q, \dot{q}, t), \qquad p = \frac{\partial L}{\partial \dot{q}}.$$

The integral (A.9) gives an expression for the propagator (the transition amplitude) of the system between the given points during the given time interval. It is a solution of the Schrödinger equation

$$i\hbar \frac{\partial}{\partial t''} U(t'', q''|t', q') = \hat{H}'' U(t'', q''|t', q') \tag{A.11}$$

with the delta-function initial condition:

$$U(t', q''|t', q') = \delta(q'' - q'). \tag{A.12}$$

The propagator is a matrix element of the evolution operator between the coordinate eigenvectors:

$$U(t'',q''|t',q') = \langle q''|U(t'',t')|q'\rangle.$$

The evolution of the system is described by the propagator (or the evolution operator) as follows:

$$\begin{aligned} |\psi_{t''}\rangle &= U(t'',t')|\psi_{t'}\rangle \\ \psi_{t''}(q'') &= \int \mathrm{d}q'\, U(t'',q''|t',q')\, \psi_{t'}(q'). \end{aligned}$$

An alternative (and often preferable) expression for a propagator is the integral over paths in phase space. Such a path [p,q] can be defined as a pair consisting of a path [q] in the space of positions (configuration space) and a path

$$[p] = \{p(t)|t' \le t \le t''\}$$

in the space of momenta. In this case the expression for a propagator is

$$U(t'',q''|t',q') = \int \mathrm{d}[p] \int \mathrm{d}[q] \exp\left(\frac{i}{\hbar}\int_{t'}^{t''} (p\dot{q} - H(p,q,t))\, \mathrm{d}t\right). \quad \text{(A.13)}$$

A3.2. DEFINITION OF A PATH INTEGRAL

Path integrals used for constructing a quantum system propagator, can be defined with the help of *discretization* or *'skeletonization'*. For an integral (A.9) this means that continuous paths $[q]$ should be replaced by continuous piecewise linear curves (broken lines) of the type drawn in Fig. A.1. The nodes of the broken line lie on the path, $q_i = q(t_i)$, $t_i = i\Delta t + t'$. Integration $d[q]$ over continuous paths $[q]$ should be replaced by integration over all possible positions of the nodes,

$$\prod_{i=1}^{N-1} dq_i.$$

This gives an approximation for a path integral. The precise value of the path integral can be obtained in the limit when the time interval

$$\Delta t = \frac{(t'' - t')}{N}$$

between nodes of broken lines tends to zero.

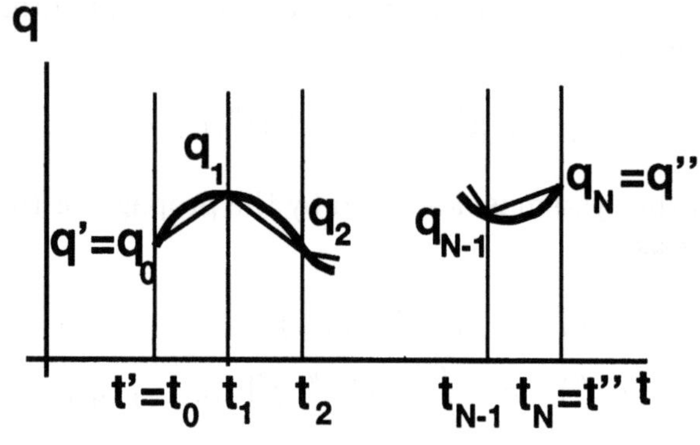

Figure A.1. Skeletonization of a path $[q]$: an arbitrary path in the configuration space is replaced by a broken line with the nodes at specified instants.

The action $S[q]$ of a system should be replaced in the process of skeletonization by the corresponding function of nodes. For the Lagrangian of the form

$$L = \frac{1}{2}m\dot{q}^2 - V(t,q)$$

one has the skeletonization of the action a:

$$S(q_0, q_1, \ldots, q_N) = \sum_{i=1}^{N} \left[\frac{1}{2}m \left(\frac{q_i - q_{i-1}}{\Delta t} \right)^2 - V(t_i, q_i) \right] \Delta t. \tag{A.14}$$

Introducing the normalizing factor

$$\prod_{i=1}^{N} \left(\frac{m}{2\pi i\hbar\Delta t} \right)^{1/2}, \tag{A.15}$$

one has, finally,

$$\begin{aligned} U(t'',q''|t',q') &= \lim_{N\to\infty} U_{(N)}(t'',q''|t',q') \qquad \text{(A.16)} \\ &= \lim_{N\to\infty} \left(\frac{m}{2\pi i\hbar\Delta t} \right)^{1/2} \int \prod_{i=1}^{N-1} \left(\frac{m}{2\pi i\hbar\Delta t} \right)^{1/2} \mathrm{d}q_i \\ &\quad \times \exp\left\{ \frac{i}{\hbar} \sum_{i=1}^{N} \left[\frac{1}{2}m\frac{(q_i - q_{i-1})^2}{\Delta t} - V(t_i, q_i)\Delta t \right] \right\}. \end{aligned}$$

It can be shown that this limit gives the solution of the problem (A.11), (A.12).

The definition (A.16) seems to be not quite satisfactory because of the arbitrary choice of the normalizing factor (A.15). The skeletonization definition of a phase space path integral (A.13) has no such shortcoming. It makes use of the same continuous piecewise linear skeletonization of paths $[q]$ in the configuration space and of the piecewise constant skeletonization of paths $[p]$ in the momentum space (Fig. A.2).

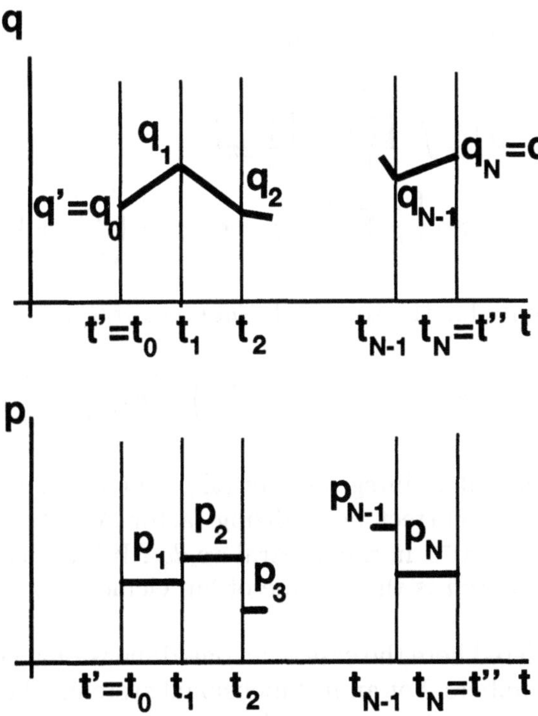

Figure A.2. Skeletonization of paths in the phase space: broken lines for $[q]$ and piecewise constant curves for $[p]$; constant values of momentum in the approximation of $[p]$ are independent of the slopes in the approximation for $[q]$.

Then the measure $\mathrm{d}[p]\mathrm{d}[q]$ should be skeletonized as

$$\prod_{i=1}^{N-1} \mathrm{d}q_i \prod_{i=1}^{N} \frac{\mathrm{d}p_i}{2\pi\hbar}, \tag{A.17}$$

and the action

$$S[p,q] = \int_{t'}^{t''} \left[p\mathrm{d}q - \left(\frac{p^2}{2m} + V(t,q) \right) \mathrm{d}t \right]$$

should be skeletonized as follows:

$$S(p_1,\ldots,p_N;q_0,q_1,\ldots,q_N)$$
$$=\sum_{i=1}^{N}\left[p_i(q_i-q_{i-1})-\left(\frac{p_i^2}{2m}+V(t_i,q_i)\right)\Delta t\right].$$

The resulting formula for the propagator is

$$\begin{aligned} U(t'',q''|t',q') &= \lim_{N\to\infty} U_{(N)}(t'',q''|t',q') \\ &= \lim_{N\to\infty}\int\prod_{i=1}^{N-1}\mathrm{d}q_i\prod_{i=1}^{N}\frac{\mathrm{d}p_i}{2\pi\hbar} \\ &\quad\times\exp\left\{\frac{i}{\hbar}\sum_{i=1}^{N}\left[p_i(q_i-q_{i-1})-\left(\frac{p_i^2}{2m}+V(t_i,q_i)\right)\Delta t\right]\right\}. \end{aligned} \tag{A.18}$$

Using the formula for the so called *Gaussian integral*

$$\int_{-\infty}^{\infty}\exp(ax^2+cx)\,\mathrm{d}x=\left(\frac{\pi}{-a}\right)^{1/2}\exp\left(-\frac{c^2}{4a}\right), \tag{A.19}$$

one can explicitly calculate integrals over all p_i in (A.18). Then the formula (A.16) arises with the correct normalizing factor. What about the trivial normalizing factor $(2\pi\hbar)^{-1}$ in the integral $dp_i dq_i$? It has an evident physical interpretation, since $2\pi\hbar$ is the volume of 'an elementary quantum cell' in phase space.

We have considered here the case of an one-dimensional system, so therefore p, q are real numbers. For an n-dimensional system, when p, q are (real) n-vectors almost all formulas are valid, but scalar products of vectors should be taken instead of products of numbers, for example,

$$p_{(i)}(q_{(i)}-q_{(i-1)})=\sum_{a=1}^{n}p_{(i)a}(q^a_{(i)}-q^a_{(i-1)}).$$

In addition to this, the normalizing factor $(2\pi\hbar)^{-1}$ in (A.18) (respectivelyly $(m/2\pi i\hbar\Delta t)^{1/2}$ in (A.16)) should be taken for each degree of freedom, leading to the corresponding measures

$$\prod_{i=1}^{N-1}\mathrm{d}^n q_i\prod_{i=1}^{N}\frac{\mathrm{d}^n p_i}{(2\pi\hbar)^n}\ \text{and}\ \left(\frac{m}{2\pi i\hbar\Delta t}\right)^{n/2}\prod_{i=1}^{N-1}\left(\frac{m}{2\pi i\hbar\Delta t}\right)^{n/2}\mathrm{d}^n q_i. \tag{A.20}$$

A3.3. GAUSSIAN PATH INTEGRALS

The path integral for a harmonic oscillator

$$H = \frac{p^2}{2m} + \frac{m\omega^2 q^2}{2} - F(t)q, \tag{A.21}$$

$$L = \frac{m\dot{q}^2}{2} - \frac{m\omega^2 q^2}{2} + F(t)q \tag{A.22}$$

is an example of a Gaussian path integral calculated with the help of the Gaussian integral (A.19). The details of the calculation may be found in (FH65) or (Men93). The result is

$$U(t'', q''|t', q') = \left(\frac{m\omega}{2\pi i\hbar \sin \omega T}\right)^{1/2} \exp\left(\frac{i}{\hbar} S[q_{\text{class}}]\right) \tag{A.23}$$

where $[q_{\text{class}}]$ is a classical trajectory between the given points q', q''.

Consider now, briefly, the whole class of Gaussian path integrals. The starting point for this consideration is an one-dimensional *Gaussian integral* (A.19). It will be convenient to rewrite it in the form

$$\int_{-\infty}^{\infty} \mathrm{d}q \exp\left(-\frac{1}{2}q^2 + cq\right) = (2\pi)^{1/2} \exp\left(\frac{1}{2}c^2\right) \tag{A.24}$$

Taking a product of n integrals of this type, one has in fact a Gaussian integral over n-dimensional vectors:

$$\int \prod_{i=1}^{n} \mathrm{d}q_i \exp\left(-\frac{1}{2}(\mathbf{q}, \mathbf{q}) + (\mathbf{c}, \mathbf{q})\right) = (2\pi)^{n/2} \exp\left(\frac{1}{2}(\mathbf{c}, \mathbf{c})\right)$$

where the scalar product is introduced:

$$(\mathbf{c}, \mathbf{q}) = \sum_{i=1}^{n} c_i q_i.$$

Substituting $A^{1/2}\mathbf{q}$ for $\mathbf{q}$ and $A^{-1/2}\mathbf{c}$ for $\mathbf{c}$ with a symmetrical matrix A and then changing variables of integration, we easily obtain

$$\int \left|\det \frac{A}{2\pi}\right|^{1/2} \prod_{i=1}^{n} \mathrm{d}q_i \exp\left(-\frac{1}{2}(\mathbf{q}, A\mathbf{q}) + (\mathbf{c}, \mathbf{q})\right) = \exp\left(\frac{1}{2}(\mathbf{c}, A^{-1}\mathbf{c})\right). \tag{A.25}$$

Consider now the path

$$[q] = \{q(t) | t' \leq t \leq t''\}$$

as a vector with an infinite number of components $q(t)$ (the argument t playing the role of an index numbering these components) and analogously for the path

$$[c] = \{c(t)|t' \le t \le t''\}.$$

Then the last formula can be rewritten as follows:

$$\int \mathrm{d}[q] \exp\left(-\frac{1}{2}([q], A[q]) + ([c], [q])\right) = \exp\left(\frac{1}{2}([c], A^{-1}[c])\right) \quad \text{(A.26)}$$

where the scalar product is defined for paths as

$$([c], [q]) = \int_{t'}^{t''} \mathrm{d}t\, c(t)q(t), \quad \text{(A.27)}$$

A is a linear operator in the space of paths, and the measure in this space is formally defined by the formula[1]

$$\mathrm{d}[q] = \left|\det \frac{A\mathrm{d}t}{2\pi}\right|^{1/2} \prod_{t=t'}^{t''} \mathrm{d}q(t). \quad \text{(A.28)}$$

The latter formula should be understood as a recipe for the procedure of skeletonization. It is not difficult to see that the earlier defined path integral (A.16) is in accordance with this recipe up to a finite number factor. But the benefit of the formula (A.26) lies in it being very easy to utilize this formula for operations with path integrals and for development of the perturbation theory.

To show this, let us take Eqs.(A.26), (A.27) as the formal definition of a *Gaussian path integral.* The concrete scheme of skeletonization expressed in Eq.(A.28) can be forgotten, because in many cases there is no need to introduce the procedure of skeletonization explicitly.

Eq.(A.26) can be used for the evaluation of path integrals with different linear operators A, including differential operators. For example, the choice

$$A = \frac{im}{\hbar}\left(\frac{\mathrm{d}^2}{\mathrm{d}t^2} + \omega^2\right), \quad c(t) = \frac{i}{\hbar}F(t) \quad \text{(A.29)}$$

converts the integral (A.26) into the path integral for a driven harmonic oscillator, but with the time integral $\int \dot{q}^2 \mathrm{d}t$ represented (with the help of integration by parts and up to a boundary term) in the form $-\int q\ddot{q}\,\mathrm{d}t$. The formula (A.26) gives then for this integral the following expression:

$$\exp\left(-\frac{i}{2\hbar}([F], A^{-1}[F])\right).$$

[1]This definition differs by a numerical factor from one accepted earlier, see below.

This proves to coincide (up to a boundary term) with $\exp\left(\frac{i}{\hbar}S[q_{\rm class}]\right)$. Thus the propagator (A.23) for a driven oscillator can be obtained with the help of Eq.(A.26) up to a numerical factor.

The difference in the numerical factors arose due to different definitions of path integral measures in both schemes of evaluation. However, in many cases a numerical factor is not essential. Moreover, a numerical factor can be found independently of the evaluation of the functional dependence of a path integral, see the spectral representation for an oscillator in (FH65; Men93).

Taking derivatives of both sides of Eq.(A.25) with respect to the components of the vector $\mathbf{c}$ shows that this is equivalent to including of a product of the corresponding components of the vector $\mathbf{q}$ in the integrand. Therefore the following formula is valid for a polynomial $\Phi(\mathbf{q})$:

$$\int \left|\det\frac{A}{2\pi}\right|^{1/2} \prod_{i=1}^{n} dq_i\ \Phi(\mathbf{q}) \exp\left(-\frac{1}{2}(\mathbf{q}, A\mathbf{q}) + (\mathbf{c},\mathbf{q})\right) = \Phi\left(\frac{\partial}{\partial \mathbf{c}}\right) \exp\left(\frac{1}{2}(\mathbf{c}, A^{-1}\mathbf{c})\right).$$

Since any function can be approximated by a polynomial to arbitrary precision, this formula is, in fact, valid for any function $\Phi(\mathbf{q})$.

Generalizing this to the case of functions (paths) one has the useful formula

$$\int d[q]\ \Phi[q] \exp\left(-\frac{1}{2}([q], A[q]) + ([c],[q])\right) = \Phi\left[\frac{\delta}{\delta c}\right] \exp\left(\frac{1}{2}([c], A^{-1}[c])\right) \tag{A.30}$$

where Φ is a functional and $\delta/\delta c(t)$ is a functional derivative. Taking $c \equiv 0$ in this formula, we obtain

$$\int d[q]\ \Phi[q] \exp\left(-\frac{1}{2}([q], A[q])\right) = \Phi\left[\frac{\delta}{\delta c}\right] \exp\left(\frac{1}{2}([c], A^{-1}[c])\right)\Big|_{c\equiv 0}. \tag{A.31}$$

This equation allows one to integrate any functional with a Gaussian measure. The result will be obtained in the form of a series, giving, in fact, a perturbation expansion for such an integral.

Finally, let us make one more remark. The Gaussian integral (A.19) converges for any complex parameter a satisfying $\mathrm{Re}\, a < 0$. Correspondingly all its generalizations, for example (A.25), (A.26), (A.30), (A.31), have a meaning for an operator A having a positive Hermitian part. For a purely anti-Hermitian operator A, as in Eq.(A.29), one should, for its correct definition, introduce a positive Hermitian part and then take the limit when this part tends to zero. In restricted path integrals such a positive Hermitian part may be present from the very beginning, so therefore convergence of RPI is better than for unrestricted Feynman path integrals (see Sect. 6.2).

References

S. Albeverio, V. N. Kolokol'tsov, and O. G. Smolyanov. Continuous quantum measurement: Local and global approaches. *Reviews in Math. Phys.*, **9**:907–920, 1997.

J. Audretsch and M. B. Mensky. Continuous fuzzy measurement of energy for a two-level system. *Phys. Rev.*, **A 56**:44–54, 1997.

J. Audretsch and M. B. Mensky. Realization scheme for continuous fuzzy measurement of energy and the monitoring of a quantum transition. 1998. quant-ph/9808062.

J. Audretsch, M. B. Mensky, and V. A. Namiot. How to visualize a quantum transition of a single atom. *Physics Letters*, **A 237**:1–9, 1997.

J. Audretsch, M. B. Mensky, and A. D. Panov. Zeno effect preventing rabi transitions onto an unstable energy level. *Physics Letters*, **A 261**:44–50, 1999.

Y. Aharonov and M. Vardi. Meaning of an individual "Feynman path". *Phys. Rev.*, **D 21**:2235–2240, 1980.

I. Bloch and D. A. Burba. Presence of a particle in a given space-time region and the continuous action of a particle detector. *Phys. Rev.*, **D 10**:3206–3218, 1974.

V. P. Belavkin. A new wave equation for a continuous nondemolition measurement. *Phys. Lett.*, **A 140**:355–358, 1989.

M. Brune, S. Haroche, V. Lefevre, J. M. Raimond, and N. Zagury. Quantum nondemolition measurement of small photon numbers by rydberg-atom phase-sensitive detection. *Phys. Rev. Lett.*, **65**:976–979, 1990.

A. Beige, G. C. Hegerfeldt, and D. G. Sondermann. *Quant. Semiclass. Opt.*, **8**:999, 1996.

V. B. Braginsky and F. Ya. Khalili. *Quantum Measurement.* Cambridge University Press, Cambridge, 1992.

P. Busch, P. Lahti, and P. Mittelstaedt. *The Quantum Theory of Measurement.* Springer Verlag, Berlin, 1991. Lecture Notes in Physics.

D. .I Blokhintsev. *Principal Questions in Quantum Mechanics.* Nauka, Moscow, 1987. in Russian.

A. Barchielli, L. Lanz, and G. M. Prosperi. A model for the macroscopic description and continual observations in quantum mechanics. *Nuovo Cimento*, **B 72**:79–121, 1982.

David Bohm. *Quantum Theory.* Prentice-Hall, inc., New York, 1952.

N. Bohr. Discussion with Einstein on epistemological problems in atomic physics. In J. A. Wheeler and W. H. Zurek, editors, *Quantum Theory and Measurement*, pages 9–49, Princeton, 1983. Princeton University Press. Originally published in "Albert Einstein: Philosopher–Scientist", P.A.Schilpp, ed.; pp. 200–241, The Library of Living Philosophers, Evanston (1949).

V. B. Braginsky, Yu. I. Vorontsov, and F. Ya. Khalili. Quantum features of a ponderomotive meter of electromagnetic energy. *Sov. Phys. JETP*, **46**:705–706, 1977.

T. Calarco. Impulsive quantum measurements — restricted path-integral vs von-Neumann collapse. *Nuovo Cimento*, **B 110**:1451–1461, 1995.

Howard Carmichael. *An Open Systems Approach to Quantum Optics.* Springer, Berlin and Heidelberg, 1993. Lectures presented at the Universite Libre de Bruxelles.

C. M. Caves. Quantum-mechanics of measurements distributed in time. *Phys. Rev.*, **D 33**:1643, 1986.

S. M. Chumakov, K.-E. Hellwig, and A. L. Rivera. Quantum Zeno effect in unitary quantum mechanics. *Physics Letters A*, 197:73–82, 1995.

A. O. Caldeira and A. J. Leggett. Path integral approach to quantum Brownian motion. *Physica*, **A 121**:587–616, 1983.

R. J. Cook. What is quantum jumps? *Phys. Scr.*, T 21:49–51, 1988.

C. B. Chiu, E. C. G. Sudarshan, and B. Misra. Time evolution of unstable quantum states and a resolution of Zeno's paradox. *Phys. Rev.*, **D 16**:520–529, 1977.

C. M. Caves, K. S. Thorne, W. P. Drever, V. D. Sandberg, and M. Zimmermann. On the measurement of a weak classical force coupled to a quantum-mechanical oscillator. *Rev. Mod. Phys.*, **52**:341–392, 1980.

E. B. Davies. *Quantum Theory of Open Systems.* Academic Press, London, New York et al., 1976.

L. Diosi, N. Gisin, J. Halliwell, and I.C.Percival. Decoherent histories and quantum state diffusion. *Phys. Rev. Lett.*, **74**:203–207, 1995.

R. H. Dicke. Interaction-free quantum measurement: A paradox? *Am. J. Phys.*, **49**:925–930, 1981.

R. H. Dicke. On observing the absence of an atom. *Found. Phys.*, **16**:107–113, 1986.

L. Diosi. Models for universal reduction of macroscopic quantum fluctuations. *Phys. Rev.*, **A 40**:1165–1174, 1989.

L. Diosi. Coarse graining and decoherence translated into von Neumann language. *Phys. Lett.*, **B 280**:71–74, 1992.

P. A. M. Dirac. *The Principles of Quantum Mechanics.* Clarendon Press, Oxford, 1958. Fourth edition.

P. A. M. Dirac. Relativity and quantum mechanics. *Fields and Quanta*, **3**:139–164, 1972.

S. Dürr, T. Nonn, and G. Rempe. Origin of quantum-mechanical complementarity probed by a 'which-way' experiment in an atom interferometer. *Nature*, **395**:33–37, 1998.

A. Einstein, B. Podolsky, and N. Rosen. Can quantum-mechanical description of physical reality be considered complete? *Phys. Rev.*, **47**:777–780, 1935.

Berthold-Georg Englert, Marlan O. Scully, and Herbert Walther. Complementarity and uncertainty. *Nature*, **375**:367–368, 1995.

A. Elitzur and L. Vaidman. Quantum mechanical interaction-free measurements. *Found. Phys.*, **23**:987–997, 1993.

H. Everett. The theory of the universal wave function. In B. S. DeWitt and Graham, editors, *The Many-Worlds Interpretation of Quantum Mechanics*, Princeton, 1973. Princeton University Press.

R. P. Feynman. Space-time approach to non-relativistic quantum mechanics. *Reviews of Modern Physics*, **20**:367, 1948.

R. P. Feynman and A. R. Hibbs. *Quantum Mechanics and Path Integrals.* McGraw-Hill Book Company, New York, 1965.

R. Feynman, R. Leighton, and M. Sands. *The Feynman Lectures on Physics.* Addison Wesley, Reading etc., 1965.

Charles N. Friedman. Continual measurements in the spacetime formation of nonrelativistic quantum mechanics. *Annals of Physics*, **98**:87–97, 1976.

R. P. Feynman and F. L. Vernon, jr. The theory of a general quantum system interacting with a linear dissipative system. *Ann. Phys. (N.Y.)*, **24**:118–173, 1963.

N. Gisin. Stochastic quantum dynamics and relativity. *Helv. Phys. Acta*, **62**:363–371, 1989.

D. Giulini, E. Joos, C. Kiefer, J. Kupsch, I.-O. Stamatescu, and H. D. Zeh. *Decoherence and the Appearance of a Classical World in Quantum Theory.* Springer, Berlin etc., 1996.

B. M. Garraway and P. L. Knight. Evolution of quantum superpositions in open environments: Quantum trajectories, jumps, and localization in phase space. *Phys. Rev.*, **A 50**:2548–2563, 1994.

G. A. Golubtsova and M. B. Mensky. Quantum nondemolition measurements from path integral. *Intern. J. of Modern Phys.*, **A 4**:2733–2750, 1989.

M. J. Gagen and G. J. Milburn. Atomic tests of the Zeno effect. *Phys. Rev.*, **A 47**:1467–1479, 1993.

M. Gell-Mann and J. B. Hartle. Quantum mechanics in the light of quantum cosmology. In W. H. Zurek, editor, *Complexity, Entropy, and the Physics of Information*, pages 425–458. Addison-Wesley, Reading, 1990.

M. Gell-Mann and J. Hartle. Classical equations for quantum systems. *Phys. Rev.*,

D 47:3345, 1993.

G. C. Ghirardi, C. Omero, T. Weber, and A. Rimini. Small-time behavior of quantum nondecay probability and Zeno's paradox in quantum mechanics. *Nuovo Cimento*, **A 52**:421–442, 1979.

R. B. Griffiths. Consistent histories and the interpretation of quantum mechanics. *J. Statist. Phys.*, **36**:219–272, 1984.

G. C. Ghirardi, A. Rimini, and T. Weber. Unified dynamics for microscopic and macroscopic systems. *Phys. Rev.*, **D 34**:470–491, 1986.

M. J. Gagen, H. M. Wiseman, and G. J. Milburn. Continuous position measurements and the quantum Zeno effect. *Phys. Rev.*, **A 48**:132, 1993.

S. Haroche, M. Brune, and J. R. Raimond. Experiments with single atoms in a cavity — entanglement, schrodingers cats and decoherence. *Phylosophical Transactions of the Royal Society of Oondon Series A*, **A 355**:2367–2380, 1997.

Carl W. Helstrom. *Quantum Detection and Estimation Theory.* Academic Press, New York etc., 1976.

A. Holevo. *Probabilistic and Statistical Aspects of Quantum Theory.* North Holland, Amsterdam, 1982.

Wendell Holladay. A simple quantum eraser. *Phys. Lett.*, **A 183**:280–282, 1993.

D. Home and M. A. B. Whitaker. A conceptual analysis of quantum Zeno; paradox, measurement, and experiment. *Annals of Physics (USA)*, **258**:237–285, 1997.

W. M. Itano, D. J. Heinzen, J. J. Bollinger, and D. J. Wineland. Quantum Zeno effect. *Phys. Rev.*, **A 41**:2295–2300, 1990.

K. Ito. *Nagoya Math. J.*, **3**:55–65, 1951.

C. Itzykson and J.-B. Zuber. *Quantum Field Theory.* MacGraw-Hill, New York, 1980.

E. Joos. Continuous measurement: Watchdog effect versus golden rule. *Phys. Rev.*, **D 29**:1626–1633, 1984.

E. Joos and H. D. Zeh. The emergence of classical properties through interaction with the environment. *Z. Phys.*, **B 59**:223–243, 1985.

M. Kac. *Probability and Related Topics in Physical Sciences.* Interscience, New York, 1958.

B. B. Kadomtsev. *Dynamics and Information.* Uspekhi Fiz. Nauk editors, Moscow, 1997.

A. Karlson, G. Björk, and E. Forsberg. "Interaction" (energy exchange) free and quantum nondemolition measurements. *Phys. Rev. Lett.*, **80**:1198–1201, 1998.

F. Ya. Khalili. *Vestnik Mosk. Universiteta, Ser. 3, Phys., Astr.*, **22**:37, 1981.

F. Ya. Khalili. Quantum Zeno paradox in measurements with finite precision. *Vestnik Mosk. Universiteta, ser. 3*, pages no.5, p.13, 1988.

J. R. Klauder. Wiener measure regularizytion for quantum mechanical path integrals. In H. Mitter and L. Pittner, editors, *Recent Development in Mathematical Physics (Proc. 26th Int. Universitätswochen Kernphys., Schladming, 17-27 February 1987)*, pages 80–99, Berlin etc., 1987. Springer-Verlag.

Hagen Kleinert. *Path Integrals in Quantum Mechanics, Statistics, and Polymer Physics.* World Scientific, Singapore, 1990.

A. Konetchnyi, M. B. Mensky, and V. Namiot. The physical model for monitoring position of a quantum particle. *Phys. Lett.*, **A 177**:283–289, 1993.

Peter Knight. Quantum optics: Practical Schrödinger cats. *Nature*, **357**:438–439, 1992.

Peter Knight. Where the weirdness comes from. *Nature*, **395**:12–13, 1998.

K. Kraus. Measuring process in quantum mechanics. i. continuous observation and the watchdog effect. *Foundation of Physics*, **11**:547–576, 1981.

Karl Kraus. *States, Effects, and Operations.* Springer-Verlag, Berlin, Heidelberg, New York and Tokyo, 1983. Lecture Notes in Physics.

G. Kwiat, Aephraim M. Steinberg, and Raymond Y. Chiao. Observation of a "quantum eraser": A revival of coherence in a two-photon interference experiment. *Phys. Rev.*, **A 45**:7729–7739, 1992.

P. Kwiat, H. Weinfurter, T. Herzog, A. Zeilinger, and M. A. Kasevich. Interaction-free measurement. *Phys. Rev. Lett.*, **74**:4763–4766, 1995.

Rolf Landauer. The physical nature of information. *Phys. Lett.*, **A 217**:188–193, 1996.
G. Lindblad. On the generators of quantum dynamical semigroups. *Commun. Math. Phys.*, **48**:119–130, 1976.
L. D. Landau and E. M. Lifshitz. *Quantum Mechanics. Nonrelativistic Theory.* Pergamon Press, Oxford, 1987.
Michael Lockwood. 'many minds' interpretations of quantum mechanics. *Brit. J. Phil. Sci.*, **47**:159–188, 1996.
G. Ludwig. *Foundations of Quantum Mechanics I.* Springer, New York, Heidelberg, Berlin, 1983.
L. I. Mandelstam. Lections on foundations of quantum mechanics. In *Complete Collection of Works*, volume V, Moscow, 1950. USSR Academy of Sciences publishers. in Russian.
L. D. Mandelstam. *Lections on Optics, Relativity and Quantum Mechanics.* Nauka, Moscow, 1972. in Russian.
M. S. Marinov. *JETP Lett.*, **15**:677, 1972.
M. S. Marinov. *Yadern. Fiz. (Sov. J. Nucl.Phys.)*, **19**:350, 1974.
M. S. Marinov. A quantum theory with possible leakage of information. *Nucl. Phys.*, **B 253**:609–620, 1985.
H. Martens. *The Uncertainty Principle.* Eindhoven, 1991. Thesis.
C. Morette-DeWitt. Feynman's path integral. definition without limiting procedure. *Commun. Math. Phys.*, **28**:47–67, 1972.
M. B. Mensky. Quantum restrictions for continuous observation of an oscillator. *Phys. Rev.*, **D 20**:384–387, 1979.
M. B. Mensky. Quantum restrictions on measurability of parameters of motion for a macroscopic oscillator. *Sov. Phys.-JETP*, **50**:667–674, 1979.
M. B. Mensky. *The Path Group: Measurements, Fields, Particles.* Nauka, Moscow, 1983. In Russian; Japanese expanded translation: Yoshioka, Kyoto, 1988.
M. B. Mensky. Evolution of a quantum system subject to continuous measurement. *Theor. Math. Phys.*, **75**:357–365, 1988.
M. B. Mensky. Some ideas of Niels Bohr in the light of quantum theory of continuous measurements. In *Niels Bohr and Physics of XX Century*, pages 203–214, Kiev, 1988. Naukova Dumka publishers. In Russian.
M. B. Mensky. The action uncertainty principle in continuous quantum measurements. *Phys. Lett.*, **A 155**:229–235, 1991.
M. B. Mensky. The action uncertainty principle and quantum gravity. *Phys. Lett.*, **A 162**:219–222, 1992.
M. B. Mensky. Continuous quantum measurements and the action uncertainty principle. *Foundations of Physics*, **22**:1173–1193, 1992.
M. B. Mensky. Difficulties in mathematical definition of path integrals are overcome in theory of quantum continuous measurements. *Teor. Mat. Fiz.*, **93**:264–272, 1992.
M. B. Mensky. *Continuous Quantum Measurements and Path Integrals.* IOP Publishing, Bristol and Philadelphia, 1993.
M. B. Mensky. Continuous quantum measurements: Restricted path integrals and master equations. *Phys. Lett.*, **A 196**:159–167, 1994.
M. B. Mensky. The Action Uncertainty Principle for continuous measurements. *Physics Letters*, **A 219**:137–144, 1996.
M. B. Mensky. Restricted path integrals for open quantum systems. In M. Namiki, I. Ohba, K. Maeda, and Y. Aizawa, editors, *Quantum Physics, Chaos Theory, and Cosmology*, pages 257–273, Woodbury and New York, 1996. American Institute of Physics.
M. B. Mensky. Decoherence in continuous measurements: from models to phenomenology. *Foundations of Physics*, **27**:1637–1654, 1997.
M. B. Mensky. Finite resolution of time in continuous measurements: Phenomenology and the model. *Phys. Lett.*, **A 231**:1–8, 1997.
M. B. Mensky. Continuously measured systems, path integrals and information. *J. Theor.*

Phys., **37**:273–280, 1998.

M. B. Mensky. Decoherence and the theory of continuous quantum measurements. *Physics-Uspekhi*, **41**:923–940, 1998. quant-ph/9812017.

M. B. Mensky. Relativistic quantum measurements, Unruh effect and black holes. *Teor. Mat. Fiz*, 115:215–232, 1998.

M. B. Mensky. Quantum Zeno effect in the decay onto an unstable level. *Phys. Letters*, **A 257**:227–231, 1999.

Albert Messiah. *Quantum Mechanics*, volume I. North-Holland Publishing Co., Amsterdam, 1970.

G. J. Milburn. Quantum measurement theory of optical heterodyne detection. *Phys. Rev.*, **A 36**:5271, 1987.

G. J. Milburn. Quantum Zeno effect and motional narrowing in a two-level system. *J. Opt. Soc. Am.*, **B 5**:1317, 1988.

S. Machida and M. Namiki. Theory of measurement in quantum mechanics. mechanism of reduction of wave packet. i. *Progr. Theor. Phys.*, **63**:1457–1473, 1980.

S. Machida and M. Namiki. Theory of measurement in quantum mechanics. mechanism of reduction of wave packet. ii. *Progr. Theor. Phys.*, **63**:1833–1847, 1980.

M. B. Mensky, R. Onofrio, and C. Presilla. Continuous quantum monitoring of position of nonlinear oscillators. *Phys. Lett.*, **A 161**:236–240, 1991.

Michael B. Mensky, Roberto Onofrio, and Carlo Presilla. Optimal monitoring of position in nonlinear quantum systems. *Phys. Rev. Lett.*, **70**:2825–2828, 1993.

B. Misra and E. C. G. Sudarshan. The Zeno's paradox in quantum theory. *J. Math. Phys.*, **18**:756–763, 1977.

H. Nakazato, M. Namiki, S. Pascazio, and H. Rauch. Understanding of the quantum Zeno effect. *Phys. Lett.*, **A 217**:203–208, 1996.

R. Omnès. From Hilbert space to common sense: A synthesis of recent progress in the interpretation of quantum mechanics. *Ann. Phys. (N.Y.)*, **201**:354–447, 1990.

R. Omnès. *The Interpretation of Quantum Mechanics.* Princeton University Press, Princeton, New Jersey, 1994.

R. Onofrio, C. Presilla, and U. Tambini. Quantum Zeno effect with the Feynman–Mensky path-integral approach. *Phys. Lett.*, **A 183**:135–140, 1993.

A. D. Panov. Quantum Zeno effect, nuclear conversion and photoionization in solids. *Annals of Physics*, **259**:1–33, 1996.

A. D. Panov. General equation for Zeno-like effects in spontaneous exponential decay. *Phys. Letters*, **A 260**:441–446, 1999.

Saverio Pascazio. Dynamical origin of the quantum Zeno effect. *Foundations of Physics*, **27**:1655–1670, 1997.

A. Peres. Zeno paradox in quantum theory. *Amer. J. Phys.*, **48**:931–932, 1980.

A. Peres. Continuous monitoring of quantum system. In *Information Complexity and Control in Quantum Physics*, pages 233–240, Wien - New York, 1985. Springer - Verlag.

A. Peres. Continuous monitoring of quantum systems. In A. Blaquiere, S. Diner, and G. Lochak, editors, *Information Complexity and Control in Quantum Physics*, pages 235–240, Wien, 1987. Springer Verlag.

Asher Peres. *Quantum Theory: Concepts and Methods.* Kluwer Academic Publishers, Dordrecht and Boston and London, 1993.

M. B. Plenio, P. L. Knight, and R. C. Thompson. Inhibition of spontaneous decay by continuous measurements. proposal for realizable experiment. *Optics Communications*, **123**:278–286, 1996.

S. Pascazio, M. Namiki, G. Badurek, and H. Rauch. Quantum Zeno effect with neutron spin. *Phys. Lett.*, **A 179**:155–160, 1993.

C. Presilla, R. Onofrio, and U. Tambini. Measurement quantum mechanics and experiments on quantum Zeno effect. *Ann. Phys. (USA)*, **248**:95–121, 1996.

J. P. Paz and W. H. Zurek. Environment-induced decoherence, classicality, and consistency of quantum histories. *Phys. Rev.*, **D 48**:2728, 1993.

Helmut Rauch. Quantum phenomena studied by neutron interferometry. *Physica Scripta*, **T 76**:24–32, 1998.

Helmut Rauch and Samuel A.. Werner. *Neutron Interferometry.* Oxford University Press, Oxford, 2000. To be published.

Rüdiger Schack, Arvid Breitenbach, and Axel Schenzle. Quantum-nondemolition measurement of small photon numbers and the preparation of number states. *Phys. Rev.*, **A 45**:3260–3267, 1992.

L. S. Schulman. *Techniques and Applications of Path Integration.* Wiley-Interscience, New York, 1981.

Marlan L. Scully and Kai Drühl. Quantum eraser: A proposed correlation experiment concerning observation and "delayed choice" in quantum mechanics. *Phys. Rev.*, **A 25**:2208–2213, 1982.

Marlan O. Scully, Berthold-Georg Englert, and Herbert Walther. Quantum optical tests of complementarity. *Nature*, **351**:111–116, 1991.

Euan Squires. *The Mystery of the Quantum World.* IOP Publishing, Bristol and Philadelphia, 1994. Second edition.

Pippa Storey, Sze Tan, Matthew Collett, and Daniel Walls. Path detection and the uncertainty principle. *Nature*, **367**:626–628, 1994.

Stig Stenholm. Simultaneous measurement of conjugate variables. *Ann. Phys. (USA)*, **218**:233–254, 1992.

Leo Stodolsky. Two state systems in media and "Turing's paradox". *Phys. Lett.*, **B 116**:464–468, 1982.

U. Tambini, C. Presilla, and R. Onofrio. Dynamics of quantum collapse in energy measurements. *Phys. Rev.*, **A 51**:967, 1995.

W. G. Unruh. Analysis of quantum-nondemolition measurements. *Phys. Rev.*, **D 18**:1764–1772, 1979.

W. G. Unruh. Maintaining coherence in quantum computers. *Phys. Rev.*, **A 51**:992, 1995.

Horst von Borzeszkowski and Michael B. Mensky. Measurability of electromagnetic field: Model and path integral methods. *Phys. Lett.*, **A 188**:249–255, 1994.

Johann von Neumann. *Mathematische Grundlagen der Quantenmechanik.* Springer, Berlin, 1932.

Johann von Neumann. *Mathematical Foundations of Quantum Mechanics.* Princeton University Press, Princeton, N.J., 1955.

D. F. Walls, M. J. Collett, and G. J. Milburn. Analysis of a quantum measurement. *Phys. Rev.*, **D 32**:3208, 1985.

Howard Wiseman and Flona Harrison. Uncertainty over complementarity? *Nature*, **377**:584, 1995.

H. M. Wiseman, F. E. Harrison, M. J. Collett, S. M. Tan, D. F. Walls, and R. B. Killip. Nonlocal momentum transfer in Welcher Weg measurements. *Phys. Rev.*, **A 56**:55–75, 1997.

E. P. Wigner. Mind-body question. In J. A. Wheeler and W. H. Zurek, editors, *Quantum Theory and Measurement*, pages 168–181, Princeton, 1983. Princeton University Press. Originally published in "The Scientist Speculates", ed. by L. G. Good, pp. 284–302, Heinemann, London (1961).

D. F. Walls and G. J. Milburn. Effect of dissipation on quantum coherence. *Phys. Rev.*, **A 31**:2403–2408, 1985.

D. F. Walls and G. J. Milburn. *Quantum Optics.* Springer-Verlag, Berlin et al., 1994.

J. A. Wheeler and W. H. Zurek, editors. *Quantum Theory and Measurement.* Princeton University Press, Princeton, 1983.

H. D. Zeh. On the interpretation of measurement in quantum theory. *Found. Phys.*, **1**:69–76, 1970.

H. D. Zeh. Toward a quantum theory of observation. *Found. Phys.*, **3**:109, 1973.

W. H. Zurek. Pointer basis of quantum apparatus: Into what mixture does the wave packet collapse? *Phys. Rev.*, **D 24**:1516, 1981.

W. H. Zurek. Environment-induced superselection rules. *Phys. Rev.*, **D 26**:1862, 1982.

Wojciech H. Zurek. Decoherence and the transition from quantum to classical. *Physics Today*, 44:36–44, 1991.

Index

Fundamental Theories of Physics

Series Editor: Alwyn van der Merwe, University of Denver, USA

1. M. Sachs: *General Relativity and Matter*. A Spinor Field Theory from Fermis to Light-Years. With a Foreword by C. Kilmister. 1982 ISBN 90-277-1381-2
2. G.H. Duffey: *A Development of Quantum Mechanics*. Based on Symmetry Considerations. 1985 ISBN 90-277-1587-4
3. S. Diner, D. Fargue, G. Lochak and F. Selleri (eds.): *The Wave-Particle Dualism*. A Tribute to Louis de Broglie on his 90th Birthday. 1984 ISBN 90-277-1664-1
4. E. Prugovečki: *Stochastic Quantum Mechanics and Quantum Spacetime*. A Consistent Unification of Relativity and Quantum Theory based on Stochastic Spaces. 1984; 2nd printing 1986 ISBN 90-277-1617-X
5. D. Hestenes and G. Sobczyk: *Clifford Algebra to Geometric Calculus*. A Unified Language for Mathematics and Physics. 1984 ISBN 90-277-1673-0; Pb (1987) 90-277-2561-6
6. P. Exner: *Open Quantum Systems and Feynman Integrals*. 1985 ISBN 90-277-1678-1
7. L. Mayants: *The Enigma of Probability and Physics*. 1984 ISBN 90-277-1674-9
8. E. Tocaci: *Relativistic Mechanics, Time and Inertia*. Translated from Romanian. Edited and with a Foreword by C.W. Kilmister. 1985 ISBN 90-277-1769-9
9. B. Bertotti, F. de Felice and A. Pascolini (eds.): *General Relativity and Gravitation*. Proceedings of the 10th International Conference (Padova, Italy, 1983). 1984 ISBN 90-277-1819-9
10. G. Tarozzi and A. van der Merwe (eds.): *Open Questions in Quantum Physics*. 1985 ISBN 90-277-1853-9
11. J.V. Narlikar and T. Padmanabhan: *Gravity, Gauge Theories and Quantum Cosmology*. 1986 ISBN 90-277-1948-9
12. G.S. Asanov: *Finsler Geometry, Relativity and Gauge Theories*. 1985 ISBN 90-277-1960-8
13. K. Namsrai: *Nonlocal Quantum Field Theory and Stochastic Quantum Mechanics*. 1986 ISBN 90-277-2001-0
14. C. Ray Smith and W.T. Grandy, Jr. (eds.): *Maximum-Entropy and Bayesian Methods in Inverse Problems*. Proceedings of the 1st and 2nd International Workshop (Laramie, Wyoming, USA). 1985 ISBN 90-277-2074-6
15. D. Hestenes: *New Foundations for Classical Mechanics*. 1986 ISBN 90-277-2090-8; Pb (1987) 90-277-2526-8
16. S.J. Prokhovnik: *Light in Einstein's Universe*. The Role of Energy in Cosmology and Relativity. 1985 ISBN 90-277-2093-2
17. Y.S. Kim and M.E. Noz: *Theory and Applications of the Poincaré Group*. 1986 ISBN 90-277-2141-6
18. M. Sachs: *Quantum Mechanics from General Relativity*. An Approximation for a Theory of Inertia. 1986 ISBN 90-277-2247-1
19. W.T. Grandy, Jr.: *Foundations of Statistical Mechanics*. Vol. I: *Equilibrium Theory*. 1987 ISBN 90-277-2489-X
20. H.-H von Borzeszkowski and H.-J. Treder: *The Meaning of Quantum Gravity*. 1988 ISBN 90-277-2518-7
21. C. Ray Smith and G.J. Erickson (eds.): *Maximum-Entropy and Bayesian Spectral Analysis and Estimation Problems*. Proceedings of the 3rd International Workshop (Laramie, Wyoming, USA, 1983). 1987 ISBN 90-277-2579-9
22. A.O. Barut and A. van der Merwe (eds.): *Selected Scientific Papers of Alfred Landé*. [*1888-1975*]. 1988 ISBN 90-277-2594-2

Fundamental Theories of Physics

23. W.T. Grandy, Jr.: *Foundations of Statistical Mechanics*. Vol. II: *Nonequilibrium Phenomena*. 1988 ISBN 90-277-2649-3
24. E.I. Bitsakis and C.A. Nicolaides (eds.): *The Concept of Probability*. Proceedings of the Delphi Conference (Delphi, Greece, 1987). 1989 ISBN 90-277-2679-5
25. A. van der Merwe, F. Selleri and G. Tarozzi (eds.): *Microphysical Reality and Quantum Formalism, Vol. 1*. Proceedings of the International Conference (Urbino, Italy, 1985). 1988 ISBN 90-277-2683-3
26. A. van der Merwe, F. Selleri and G. Tarozzi (eds.): *Microphysical Reality and Quantum Formalism, Vol. 2*. Proceedings of the International Conference (Urbino, Italy, 1985). 1988 ISBN 90-277-2684-1
27. I.D. Novikov and V.P. Frolov: *Physics of Black Holes*. 1989 ISBN 90-277-2685-X
28. G. Tarozzi and A. van der Merwe (eds.): *The Nature of Quantum Paradoxes*. Italian Studies in the Foundations and Philosophy of Modern Physics. 1988 ISBN 90-277-2703-1
29. B.R. Iyer, N. Mukunda and C.V. Vishveshwara (eds.): *Gravitation, Gauge Theories and the Early Universe*. 1989 ISBN 90-277-2710-4
30. H. Mark and L. Wood (eds.): *Energy in Physics, War and Peace*. A Festschrift celebrating Edward Teller's 80th Birthday. 1988 ISBN 90-277-2775-9
31. G.J. Erickson and C.R. Smith (eds.): *Maximum-Entropy and Bayesian Methods in Science and Engineering*. Vol. I: *Foundations*. 1988 ISBN 90-277-2793-7
32. G.J. Erickson and C.R. Smith (eds.): *Maximum-Entropy and Bayesian Methods in Science and Engineering*. Vol. II: *Applications*. 1988 ISBN 90-277-2794-5
33. M.E. Noz and Y.S. Kim (eds.): *Special Relativity and Quantum Theory*. A Collection of Papers on the Poincaré Group. 1988 ISBN 90-277-2799-6
34. I.Yu. Kobzarev and Yu.I. Manin: *Elementary Particles. Mathematics, Physics and Philosophy*. 1989 ISBN 0-7923-0098-X
35. F. Selleri: *Quantum Paradoxes and Physical Reality*. 1990 ISBN 0-7923-0253-2
36. J. Skilling (ed.): *Maximum-Entropy and Bayesian Methods*. Proceedings of the 8th International Workshop (Cambridge, UK, 1988). 1989 ISBN 0-7923-0224-9
37. M. Kafatos (ed.): *Bell's Theorem, Quantum Theory and Conceptions of the Universe*. 1989 ISBN 0-7923-0496-9
38. Yu.A. Izyumov and V.N. Syromyatnikov: *Phase Transitions and Crystal Symmetry*. 1990 ISBN 0-7923-0542-6
39. P.F. Fougère (ed.): *Maximum-Entropy and Bayesian Methods*. Proceedings of the 9th International Workshop (Dartmouth, Massachusetts, USA, 1989). 1990 ISBN 0-7923-0928-6
40. L. de Broglie: *Heisenberg's Uncertainties and the Probabilistic Interpretation of Wave Mechanics*. With Critical Notes of the Author. 1990 ISBN 0-7923-0929-4
41. W.T. Grandy, Jr.: *Relativistic Quantum Mechanics of Leptons and Fields*. 1991 ISBN 0-7923-1049-7
42. Yu.L. Klimontovich: *Turbulent Motion and the Structure of Chaos*. A New Approach to the Statistical Theory of Open Systems. 1991 ISBN 0-7923-1114-0
43. W.T. Grandy, Jr. and L.H. Schick (eds.): *Maximum-Entropy and Bayesian Methods*. Proceedings of the 10th International Workshop (Laramie, Wyoming, USA, 1990). 1991 ISBN 0-7923-1140-X
44. P. Pták and S. Pulmannová: *Orthomodular Structures as Quantum Logics*. Intrinsic Properties, State Space and Probabilistic Topics. 1991 ISBN 0-7923-1207-4
45. D. Hestenes and A. Weingartshofer (eds.): *The Electron*. New Theory and Experiment. 1991 ISBN 0-7923-1356-9

Fundamental Theories of Physics

46. P.P.J.M. Schram: *Kinetic Theory of Gases and Plasmas.* 1991 ISBN 0-7923-1392-5
47. A. Micali, R. Boudet and J. Helmstetter (eds.): *Clifford Algebras and their Applications in Mathematical Physics.* 1992 ISBN 0-7923-1623-1
48. E. Prugovečki: *Quantum Geometry.* A Framework for Quantum General Relativity. 1992 ISBN 0-7923-1640-1
49. M.H. Mac Gregor: *The Enigmatic Electron.* 1992 ISBN 0-7923-1982-6
50. C.R. Smith, G.J. Erickson and P.O. Neudorfer (eds.): *Maximum Entropy and Bayesian Methods.* Proceedings of the 11th International Workshop (Seattle, 1991). 1993 ISBN 0-7923-2031-X
51. D.J. Hoekzema: *The Quantum Labyrinth.* 1993 ISBN 0-7923-2066-2
52. Z. Oziewicz, B. Jancewicz and A. Borowiec (eds.): *Spinors, Twistors, Clifford Algebras and Quantum Deformations.* Proceedings of the Second Max Born Symposium (Wrocław, Poland, 1992). 1993 ISBN 0-7923-2251-7
53. A. Mohammad-Djafari and G. Demoment (eds.): *Maximum Entropy and Bayesian Methods.* Proceedings of the 12th International Workshop (Paris, France, 1992). 1993 ISBN 0-7923-2280-0
54. M. Riesz: *Clifford Numbers and Spinors* with Riesz' Private Lectures to E. Folke Bolinder and a Historical Review by Pertti Lounesto. E.F. Bolinder and P. Lounesto (eds.). 1993 ISBN 0-7923-2299-1
55. F. Brackx, R. Delanghe and H. Serras (eds.): *Clifford Algebras and their Applications in Mathematical Physics.* Proceedings of the Third Conference (Deinze, 1993) 1993 ISBN 0-7923-2347-5
56. J.R. Fanchi: *Parametrized Relativistic Quantum Theory.* 1993 ISBN 0-7923-2376-9
57. A. Peres: *Quantum Theory: Concepts and Methods.* 1993 ISBN 0-7923-2549-4
58. P.L. Antonelli, R.S. Ingarden and M. Matsumoto: *The Theory of Sprays and Finsler Spaces with Applications in Physics and Biology.* 1993 ISBN 0-7923-2577-X
59. R. Miron and M. Anastasiei: *The Geometry of Lagrange Spaces: Theory and Applications.* 1994 ISBN 0-7923-2591-5
60. G. Adomian: *Solving Frontier Problems of Physics: The Decomposition Method.* 1994 ISBN 0-7923-2644-X
61. B.S. Kerner and V.V. Osipov: *Autosolitons.* A New Approach to Problems of Self-Organization and Turbulence. 1994 ISBN 0-7923-2816-7
62. G.R. Heidbreder (ed.): *Maximum Entropy and Bayesian Methods.* Proceedings of the 13th International Workshop (Santa Barbara, USA, 1993) 1996 ISBN 0-7923-2851-5
63. J. Peřina, Z. Hradil and B. Jurčo: *Quantum Optics and Fundamentals of Physics.* 1994 ISBN 0-7923-3000-5
64. M. Evans and J.-P. Vigier: *The Enigmatic Photon.* Volume 1: The Field $\boldsymbol{B}^{(3)}$. 1994 ISBN 0-7923-3049-8
65. C.K. Raju: *Time: Towards a Constistent Theory.* 1994 ISBN 0-7923-3103-6
66. A.K.T. Assis: *Weber's Electrodynamics.* 1994 ISBN 0-7923-3137-0
67. Yu. L. Klimontovich: *Statistical Theory of Open Systems.* Volume 1: A Unified Approach to Kinetic Description of Processes in Active Systems. 1995 ISBN 0-7923-3199-0; Pb: ISBN 0-7923-3242-3
68. M. Evans and J.-P. Vigier: *The Enigmatic Photon.* Volume 2: Non-Abelian Electrodynamics. 1995 ISBN 0-7923-3288-1
69. G. Esposito: *Complex General Relativity.* 1995 ISBN 0-7923-3340-3

Fundamental Theories of Physics

70. J. Skilling and S. Sibisi (eds.): *Maximum Entropy and Bayesian Methods.* Proceedings of the Fourteenth International Workshop on Maximum Entropy and Bayesian Methods. 1996 ISBN 0-7923-3452-3
71. C. Garola and A. Rossi (eds.): *The Foundations of Quantum Mechanics Historical Analysis and Open Questions.* 1995 ISBN 0-7923-3480-9
72. A. Peres: *Quantum Theory: Concepts and Methods.* 1995 (see for hardback edition, Vol. 57) ISBN Pb 0-7923-3632-1
73. M. Ferrero and A. van der Merwe (eds.): *Fundamental Problems in Quantum Physics.* 1995 ISBN 0-7923-3670-4
74. F.E. Schroeck, Jr.: *Quantum Mechanics on Phase Space.* 1996 ISBN 0-7923-3794-8
75. L. de la Peña and A.M. Cetto: *The Quantum Dice.* An Introduction to Stochastic Electrodynamics. 1996 ISBN 0-7923-3818-9
76. P.L. Antonelli and R. Miron (eds.): *Lagrange and Finsler Geometry.* Applications to Physics and Biology. 1996 ISBN 0-7923-3873-1
77. M.W. Evans, J.-P. Vigier, S. Roy and S. Jeffers: *The Enigmatic Photon.* Volume 3: Theory and Practice of the $\boldsymbol{B}^{(3)}$ Field. 1996 ISBN 0-7923-4044-2
78. W.G.V. Rosser: *Interpretation of Classical Electromagnetism.* 1996 ISBN 0-7923-4187-2
79. K.M. Hanson and R.N. Silver (eds.): *Maximum Entropy and Bayesian Methods.* 1996 ISBN 0-7923-4311-5
80. S. Jeffers, S. Roy, J.-P. Vigier and G. Hunter (eds.): *The Present Status of the Quantum Theory of Light.* Proceedings of a Symposium in Honour of Jean-Pierre Vigier. 1997 ISBN 0-7923-4337-9
81. M. Ferrero and A. van der Merwe (eds.): *New Developments on Fundamental Problems in Quantum Physics.* 1997 ISBN 0-7923-4374-3
82. R. Miron: *The Geometry of Higher-Order Lagrange Spaces.* Applications to Mechanics and Physics. 1997 ISBN 0-7923-4393-X
83. T. Hakioğlu and A.S. Shumovsky (eds.): *Quantum Optics and the Spectroscopy of Solids.* Concepts and Advances. 1997 ISBN 0-7923-4414-6
84. A. Sitenko and V. Tartakovskii: *Theory of Nucleus.* Nuclear Structure and Nuclear Interaction. 1997 ISBN 0-7923-4423-5
85. G. Esposito, A.Yu. Kamenshchik and G. Pollifrone: *Euclidean Quantum Gravity on Manifolds with Boundary.* 1997 ISBN 0-7923-4472-3
86. R.S. Ingarden, A. Kossakowski and M. Ohya: *Information Dynamics and Open Systems.* Classical and Quantum Approach. 1997 ISBN 0-7923-4473-1
87. K. Nakamura: *Quantum versus Chaos.* Questions Emerging from Mesoscopic Cosmos. 1997 ISBN 0-7923-4557-6
88. B.R. Iyer and C.V. Vishveshwara (eds.): *Geometry, Fields and Cosmology.* Techniques and Applications. 1997 ISBN 0-7923-4725-0
89. G.A. Martynov: *Classical Statistical Mechanics.* 1997 ISBN 0-7923-4774-9
90. M.W. Evans, J.-P. Vigier, S. Roy and G. Hunter (eds.): *The Enigmatic Photon.* Volume 4: New Directions. 1998 ISBN 0-7923-4826-5
91. M. Rédei: *Quantum Logic in Algebraic Approach.* 1998 ISBN 0-7923-4903-2
92. S. Roy: *Statistical Geometry and Applications to Microphysics and Cosmology.* 1998 ISBN 0-7923-4907-5
93. B.C. Eu: *Nonequilibrium Statistical Mechanics.* Ensembled Method. 1998 ISBN 0-7923-4980-6

Fundamental Theories of Physics

94. V. Dietrich, K. Habetha and G. Jank (eds.): *Clifford Algebras and Their Application in Mathematical Physics.* Aachen 1996. 1998 ISBN 0-7923-5037-5
95. J.P. Blaizot, X. Campi and M. Ploszajczak (eds.): *Nuclear Matter in Different Phases and Transitions.* 1999 ISBN 0-7923-5660-8
96. V.P. Frolov and I.D. Novikov: *Black Hole Physics.* Basic Concepts and New Developments. 1998 ISBN 0-7923-5145-2; PB 0-7923-5146
97. G. Hunter, S. Jeffers and J-P. Vigier (eds.): *Causality and Locality in Modern Physics.* 1998 ISBN 0-7923-5227-0
98. G.J. Erickson, J.T. Rychert and C.R. Smith (eds.): *Maximum Entropy and Bayesian Methods.* 1998 ISBN 0-7923-5047-2
99. D. Hestenes: *New Foundations for Classical Mechanics (Second Edition).* 1999 ISBN 0-7923-5302-1; PB ISBN 0-7923-5514-8
100. B.R. Iyer and B. Bhawal (eds.): *Black Holes, Gravitational Radiation and the Universe.* Essays in Honor of C. V. Vishveshwara. 1999 ISBN 0-7923-5308-0
101. P.L. Antonelli and T.J. Zastawniak: *Fundamentals of Finslerian Diffusion with Applications.* 1998 ISBN 0-7923-5511-3
102. H. Atmanspacher, A. Amann and U. Müller-Herold: *On Quanta, Mind and Matter Hans Primas in Context.* 1999 ISBN 0-7923-5696-9
103. M.A. Trump and W.C. Schieve: *Classical Relativistic Many-Body Dynamics.* 1999 ISBN 0-7923-5737-X
104. A.I. Maimistov and A.M. Basharov: *Nonlinear Optical Waves.* 1999 ISBN 0-7923-5752-3
105. W. von der Linden, V. Dose, R. Fischer and R. Preuss (eds.): *Maximum Entropy and Bayesian Methods Garching, Germany 1998.* 1999 ISBN 0-7923-5766-3
106. M.W. Evans: *The Enigmatic Photon Volume 5: O(3) Electrodynamics.* 1999 ISBN 0-7923-5792-2
107. G.N. Afanasiev: *Topological Effects in Quantum* Mechanics 1999 ISBN 0-7923-5800-7
108. V. Devanathan: *Angular Momentum Techniques in Quantum Mechanics.* 1999 ISBN 0-7923-5866-X
109. P.L. Antonelli (ed.): *Finslerian Geometries A Meeting of Minds.* 1999 ISBN 0-7923-6115-6
110. M.B. Mensky: *Quantum Measurements and Decoherence* Models and Phenomenology. 2000 ISBN 0-7923-6227-6
111. B. Coecke, D. Moore and A. Wilce (eds.): *Current Research in Operation Quantum Logic.* Algebras, Categories, Languages. 2000 ISBN 0-7923-6258-6

KLUWER ACADEMIC PUBLISHERS – DORDRECHT / BOSTON / LONDON

MIX
Papier aus verantwortungsvollen Quellen
Paper from responsible sources
FSC® C105338

If you have any concerns about our products,
you can contact us on
ProductSafety@springernature.com

In case Publisher is established outside the EU,
the EU authorized representative is:
Springer Nature Customer Service Center GmbH
Europaplatz 3, 69115 Heidelberg, Germany

Printed by Libri Plureos GmbH
in Hamburg, Germany